SEARCHING THE SKIES

THE LEGACY OF THE UNITED STATES COLD WAR DEFENSE RADAR PROGRAM

HEADQUARTERS AIR COMBAT COMMAND
JUNE 1997

Library of Congress Cataloging-in-Publication Data

Winkler, David F. (David Frank), 1958–
    Searching the skies : the legacy of the United States Cold War defense radar program / [by David F. Winkler].
       p. cm.
    Prepared for United States Air Force Headquarters Air Combat Command.
    Includes bibliographical references (p.   ) and index.
    1. Radar defense networks—United States—History. 2. Cold War. I. United States. Air Force. Air Combat Command. II. Title.
UG612.3.W56 1997
355.4—dc21   97-20912

United States Air Force Air Combat Command, Langley AFB, VA
U.S. Army Construction Engineering Research Laboratories, Champaign, IL

CONTRACT PARTICULARS

This Study Was Prepared for
**United States Air Force
Headquarters Air Combat Command**

By

David F. Winkler

Under the supervision of
**Julie L. Webster
Principal Investigator**

**United States Army Construction Engineering Research Laboratories
Champaign, Illinois**

Under
Reimbursable Order No. ND94-0416

June 1997

**United States Air Force
Air Combat Command**

Contents

Part I: The History of Defense Radar Programs — 1

Introduction — 3

1   The Evolution of Air Defense (1918–1959) — 7
Early Development of Air Defense — 7
The Post-War Era — 14
Building the Network — 22
The Debate — 24
Improving Command and Control — 29
Improving the Radar Network — 33

2   The Evolution to Aerospace Defense (1959–1979) — 37
The Radar Network After Sputnik — 39
Missile Detection and Defense — 48

3   Air Defense Revitalized (1979–1994) — 57
Looking to the Future — 57
Rebuilding the Network — 57

4   Epilogue — 63

Endnotes — 65

Part II: Systems Overview — 71

Radar Systems Classification Methods — 73
World War II Radars — 73
Early Cold War Search Radars — 76
SAGE System Compatible Search Radars — 77

SAGE System Compatible Height-finder Radars	81
Gap-Filler Radars	82
North Warning System Radars	84
Ballistic Missile Early Warning System (BMEWS) Radars	84
Federal Aviation Administration (FAA) Radars	88

Command and Control Systems   89

Semi-Automatic Ground Environment (SAGE) System	89
Backup Interceptor Control (BUIC) System	89
Joint Surveillance System (JSS)	89

Part III: Site Listings   91

Appendix: Notes for Site Listings   171

Bibliography   173

Defense Radar Acronyms   185

Index   187

FOREWORD

The Department of Defense (DoD) Legacy Resource Management Program was established under the Defense Appropriations Act of 1991 to "determine how to better integrate the conservation of irreplaceable biological, cultural, and geophysical resources with the dynamic requirements of military missions." One of Legacy's nine task areas was the Cold War Project, which seeks to "inventory, protect, and conserve [DoD's] physical and literary property and relics" associated with the Cold War.

In 1993, Dr. Paul Green, Cultural Resource Manager for the Air Force's Air Combat Command, became involved in a series of Cold War-era cultural resource initiatives. These efforts, led by the Cold War Task Area Manager for Legacy, Dr. Rebecca Hancock Cameron, ranged from site-specific documentation to national theme and context studies. For the latter, an extensive list of possible topics was developed. After much consideration, this list was honed to include only those that played a primary role in the Cold War.

During the Cold War the United States Air Force assumed a burden for the defense of the skies over the United States. To coordinate this defense, a detection and command and control system had to be put in place that involved the construction of hundreds of radar stations and command facilities across the North American continent. As the Soviet threat evolved from bombers to missiles, America's detection and command and control systems also grew more sophisticated.

Searching the Skies: The Legacy of the United States Cold War Defense Radar Program is the product of an effort by personnel at the U.S. Army Construction Engineering Research Laboratories (USACERL), working in cooperation with Dr. Green, Dr. Cameron, and other members of the military history community. The purpose of this study is to provide a tool for military installation cultural resource managers tasked with the proper stewardship of their historic and cultural resources. The study provides the basis for identifying, understanding, and evaluating DoD radar facilities associated with the Cold War.

Several members of the USACERL staff made contributions to this publication. Mr. David F. Winkler, a doctoral candidate at the American University in Washington, DC, and a Commander in the U.S. Naval Reserve, prepared the narrative. He also edited the manuscript, selected the illustrations, and compiled the bibliography. Mr. Winkler was assisted by Dr. John Lonnquest, who offered general guidance and wrote the section on Havre Air Force Station in Part III. In addition, Ms. Gloria J. Wienke served as managing editor for the project.

Julie L. Webster, RA
Principal Investigator
USACERL

PREFACE

Over the course of this project I received help from a great many people. First and foremost, I would like to thank Dr. Paul Green, the Cultural Resource Manager for the Air Force Air Combat Command, Langley Air Force Base, Virginia. Dr. Green set the scope, content, and format of this study and directed me to valuable source materials. Also, Dr. Rebecca Cameron, Department of Defense Cold War Task Area Manager for Legacy, provided additional encouragement and focus. Ms. Virge Jenkins Temme, the U.S. Army Construction Engineering Research Laboratories principal investigator at the outset of this effort, provided a valuable critique of this work. Ms. Temme's tasks as principal investigator were assumed by Ms. Julie L. Webster in May 1996. During the ensuing year of managing the project, Ms. Webster spent countless hours on administrative and technical details required to shepherd this project to completion. Her time and talents are gratefully appreciated. The text editor was Sharlyn A. Dimick. Gloria J. Wienke, USACERL managing editor, completed the final editing and packaging of this book. My colleague John Lonnquest assisted with some of the research and also provided good ideas on how to structure and clarify the narrative. He also wrote the short pieces on Havre Air Force Station in Part III of this book.

Several other historians helped along the way. At the Federal Aviation Administration, Dr. Edmund R. Preston provided access to files detailing DoD-FAA cooperation. Air Combat Command historian Dr. Paul E. McAllister provided access to Air Defense Command historical studies. Maps in the narrative are derived from these studies. Dr. McAllister also helped declassify some materials. Dr. Thomas E. Fuller at NORAD, provided additional documents and critiqued an earlier draft.

I am especially indebted to Mr. Stephen L. Johnston. A retired radar expert, Mr. Johnston provided technical critique and direction to obtain additional material for the bibliography. Mandy Whorton of Argonne National Laboratory provided valuable feedback on the Ballistic Missile Early Warning System.

Jeffrey P. Buchheit, Director of the Historical Electronics Museum near Baltimore, helped locate primary source materials and photographs used in this book. Mindy Rosewitz, Curator at the U.S. Army Communications Electronics Command Museum at Fort Monmouth, New Jersey, provided me with access to early Army Signal Corps photos and reports. Frederik Nebeker and Andrew Goldstein at the Center for the History of Electrical Engineering provided several bibliographic sources. Mary Kuykendall of the Hall of History in Schenectady, New York, provided a description of archival holdings.

Finally, acknowledgment must go to those reference librarians who seem to have a knack for locating the document containing the golden nugget. At the U.S. Air Force

Searching the Skies: The Legacy of the United States Cold War Defense Radar Program

Museum at Wright-Patterson Air Force Base, Ohio, Dave Menard pointed me to old Air Defense Command files that provided excellent background material. At Bolling Air Force Base, Washington, DC, Yvonne Kinkaid helped identify periodical sources and other studies that broadened the bibliography. Again, my thanks to all the people who helped me in this endeavor.

David F. Winkler, May 1997

Part I

The History of Defense Radar Programs

Introduction

The Air Defense Command (ADC) was formed in 1946, and since that time the effort to detect intruding objects approaching the skies and the heavens above the North American continent has been ongoing with increasing sophistication. Initially ADC used radar sets left over from World War II in a hasty program appropriately dubbed "Lashup." Lashup provided some air search protection for such important locations as the Atomic Energy Commission at Hanford, Washington. In the early 1950s, the Air Force constructed a permanent network. This network employed the first generation of post-WWII search and height-finding radars. During the late 1950s, the AN/FPS-20 search radar became a mainstay at many radar sites until the end of the Cold War. Later modifications of the AN/FPS-20 included the FPS-64/65/66/67 and 91A. Among the sites that received these radars were over 100 stations of the mobile program. While initially intended for rapid installation and dismantling, mobile radar stations were in reality hardly mobile. Eventually the mobile moniker was dropped. Because of the threat of jamming, frequency-diversity radars were designed and deployed at both permanent and mobile stations during the 1960s.

As additional radar sites were built within the continental United States, the Distant Early Warning (DEW) Line and additional radar fences were constructed in Canada and Alaska to provide the North American air defense organization early warnings of a Soviet bomber attack. In the late 1950s, this air defense organization became binational and has since been known as NORAD (the North American Air Defense Command).

While radar sensors improved, so did the nation's ability to command and control its air defenses. Using computers to accelerate the transmission and display of tracking data, Semi-Automatic Ground Environment (SAGE) centers could quickly deploy interceptors and designate missile batteries (sites) to engage hostile aircraft. During the 1960s, SAGE centers were augmented by the Backup Interceptor Control (BUIC) system.

With the Soviet launch of Sputnik on October 4, 1957, a crash effort was made to build long-range Ballistic Missile Early Warning System (BMEWS) radars to give warning of a Soviet missile launch and prevent a nuclear Pearl Harbor. At Cheyenne Mountain, Colorado, a hardened NORAD command center was carved out of granite. The center received and evaluated intelligence from sensors and other sources for use by National Command Authorities. Additional sensors, built in the 1960s, 1970s, and 1980s, gave NORAD warnings of submarine-launched ballistic missiles and objects in orbit above the United States.

Because of the missile threat, vulnerable air defense radar sites and SAGE command centers were systematically shut down and improvements were postponed. The 1970s represented a nadir in U.S. air defense capability as represented by the eventual disestablishment of the Army Air Defense Command in 1975 and the absorption of elements of the Air Force Aerospace Defense Command by Strategic Air Command (SAC) and Tactical Air Command (TAC) in 1979.

By the end of the Cold War, America's air defense capabilities had been revitalized. Technology played a key role in providing the nation with a far less manpower-intensive yet more capable detection system. SAGE command and control systems were retired in favor of Region Operation Control Centers (ROCCs) that were tied to Airborne Warning and Control System (AWACS) aircraft. Radars in Alaska were upgraded. The DEW Line was retired and replaced by the North Warning System. In the lower forty-eight states, the Air Force and Federal Aviation Administration cooperated with the Joint Surveillance System (JSS) and construction was completed on east and west coast Over-the-Horizon-Backscatter (OTH-B) radar receiver and transmitter sites.

As a result of these many programs to guard against Soviet nuclear attack, thousands of structures were built throughout North America. Antenna towers, operations centers, administration offices, and housing and utilities structures could be found at hundreds of military installations built throughout the United States and Canada from the late 1940s into the 1990s.

Although the Cold War has ended, many of these structures still exist to provide intrusion warnings over the skies of North America, to monitor and control the thousands of commercial flights that fly the continental air routes daily, and to track objects in orbit over the western hemisphere. However, as technology evolved during the Cold War and the Soviet threat diminished, many more of these facilities were abandoned or converted to other uses.

Gibbsboro Air Force Station, New Jersey, 1995. (U.S. Army Construction Engineering Research Laboratories photograph.)

Introduction

    This book provides the historical context for cultural resource managers to evaluate the significance of radar and command and control facilities. The book is divided into three parts. Part I contains a chronological narrative tracing the evolution of air and aerospace detection, command, and control during the Cold War. Part II provides an overview of key systems related to detection, command, and control. Part III provides a state-by-state listing of detection, command, and control sites. The index contains names, places, and subjects from Part I only; technical information on the systems can be found in Parts II and III.

Chapter 1

The Evolution of Air Defense (1918–1959)

During President John F. Kennedy's October 1962 confrontation with the Soviet Union over Soviet missiles in Cuba, the United States had military advantages. One advantage was the ability to defend against a possible bomber attack from the Soviet Union. Such an attack would have been detected early by a string of radar stations that stretched across northern Canada and Alaska. Once the intruding bombers crossed into North American airspace, their progress would have been monitored by an extensive system of long-range and short-range radar stations that dotted the landscape of the United States and Canada. These radar sites provided data to a system of combat and direction centers that formed the heart of what was called the Semi-Automatic Ground Environment (SAGE) system. From SAGE combat and direction centers, orders to engage the enemy bombers could be issued to the numerous interceptor squadrons, and to scores of Air Force BOMARC and Army Nike surface-to-air missile batteries scattered around the country.

However, the construction of this extensive air defense network to counter the Soviet threat had not been assured. During the late 1940s and 1950s, funding for radars, command and control centers, and interceptor aircraft and missiles often faced severe challenges from within the military, the Executive Branch, and in Congress.

Air defense concepts received little support because of a heavily institutionalized bias in the Army Air Forces that favored offense as the best defense. Prior to WWII air power advocates considered strategic bombing to be key to breaking enemy production capacity and civilian morale. The Army Air Forces pursued this doctrine with vigor over German and Japanese skies. During the Cold War, this philosophy was heralded by SAC. This chapter will trace the struggle of the proponents of defensive and offensive strategies before and during the early years of the Cold War.

Early Development of Air Defense

The Air Force based its offensive philosophy on experience, dating from World War I, that repeatedly demonstrated that defensive forces were at a disadvantage against fast, high-flying bombers. To blunt German bomber attacks against their home island during 1918, the British established an elaborate system of barrage zones around London that included searchlights, observation posts, and antiaircraft artillery. Additional defensive measures included tethering large balloons on cables to thwart low-flying bombers and imposing blackouts on British cities to complicate targeting. However, only through the introduction of many fighter squadrons recalled from France were the British able to contain the German threat. After World War I, air power advocates such as American

Searching the Skies: The Legacy of the United States Cold War Defense Radar Program

The first practical means of detecting airplanes at a distance was by listening for their noise with the aid of horns. Larger horns increased the range of detection. One horn pair funneled sound into the ears of the azimuth operator; the other pair was used by the elevation operator. (U.S. Army Signal Corps photographs.)

The Evolution of Air Defense (1918–1959)

General Billy Mitchell and Italian theorist Giulio Douchet argued that strategic bombardment would revolutionize warfare. In 1929, many Americans were impressed with the offensive advantage of the bombers in maneuvers over Ohio. There they observed that pursuit planes had difficulty approaching the faster and higher flying B-10.[1]

Not everyone in the military was convinced of the impunity of bomber aircraft. During the 1930s, Captains Claire L. Chennault and Gordon P. Saville theorized that improved pursuit aircraft using defensive tactics could challenge the bomber claim to air supremacy. A breakthrough was achieved in 1935 when the Army experimented with a Ground Control Interception (GCI) system. A brainchild of Saville, the GCI system used high-frequency radios to vector interceptor aircraft towards incoming formations of bombers that had been spotted and reported by ground observers. Although the system could not stop a determined attack, it could hinder the attackers' destruction of the target and cost the enemy an unacceptable loss of bomber aircraft.

While the Army worked to improve the GCI system and pursuit aircraft, experiments initiated by the Naval Research Laboratory in Washington, DC, and the Army Signal Corps Laboratories at Fort Monmouth, New Jersey, proceeded to develop detection devices using radio waves. The Army also attempted to develop alternative detection devices targeting aircraft engine noise and heat signatures. However, aural and thermal detection devices had only limited potential because of the difficulty in detecting sound and heat at great distances. Fortunately, Army and Navy researchers progressed in their radio wave experiments. On December 14, 1936, the Army successfully tested a pulse radar at Princeton Junction, New Jersey. They were able to bounce radio waves off an aircraft out to a range of seven miles. A series of demonstrations conducted near Fort Monmouth, New Jersey, the following May proved even more promising and impressed Secretary of War, Harry A. Woodring enough to increase research funding. Prototype radars tested in May 1937 were the direct predecessors of the Army SCR-268, SCR-270, and SCR-271 search radars.[2] Private research organizations such as Bell Telephone Laboratories joined the effort. The Navy also had success in developing radar units for deployment at sea. Tests during the late 1930s proved encouraging and the first production units joined the fleet in 1940. In an effort to name the new devices, the Navy invented the acronym RADAR for "Radio Detection And Ranging."[3] The Army also began to deploy radar sets. One unit, an SCR-270B, was operational on the morning of December 7, 1941, at Kahuku Point on Oahu, Hawaii. Unfortunately, the big sighting report called in to Pearl Harbor by the two assigned radarmen was ignored.[4]

The Battle of Britain demonstrated the validity of Chennault and Saville's theory that improved pursuit aircraft challenge bomber supremacy. In coordinating the defense of their country, the British used a radar technology that they had been developing during the 1930s. Forewarned of a German bomber attack by primitive radar sets, ground control command posts efficiently deployed Royal Air Force Hurricane and Spitfire interceptor aircraft to wreak havoc on the attackers. Although the British defenders did not prevent the Nazi attackers from bombing their targets, they exacted enough in German aircraft losses to force Germany to reconsider its bombing strategy.[5]

Searching the Skies: The Legacy of the United States Cold War Defense Radar Program

Experiments in 1937 using antennas (upper photograph) proved successful and lead to the development of the Army's first production radar—the SCR-268 (lower photograph).
(U.S. Army Signal Corps photographs.)

In 1940, as fighting continued over the skies of Britain, the Army established the ADC at Mitchel Field on Long Island, New York. Established to test systems and formulate doctrine, ADC's mission took on a sense of urgency after the Japanese attack at Pearl Harbor. By attacking the Hawaiian Islands with carrier-borne aircraft, the Japanese demonstrated that the American west coast was also vulnerable to carrier aircraft. In the wake of the attack, ADC scrambled to deploy interceptor squadrons, radar sets, and gun emplacements. Ninety-five radar stations were eventually set up along the east and west coasts using SCR-270 (mobile) and SCR-271 (fixed) radar sets. These radars had an optimum range of 150 miles distance at 20,000 feet elevation. Thousands of civilian volunteers joined Ground Observer Corps to scan the skies up and down the east and west coasts for enemy aircraft.[6]

The Evolution of Air Defense (1918–1959)

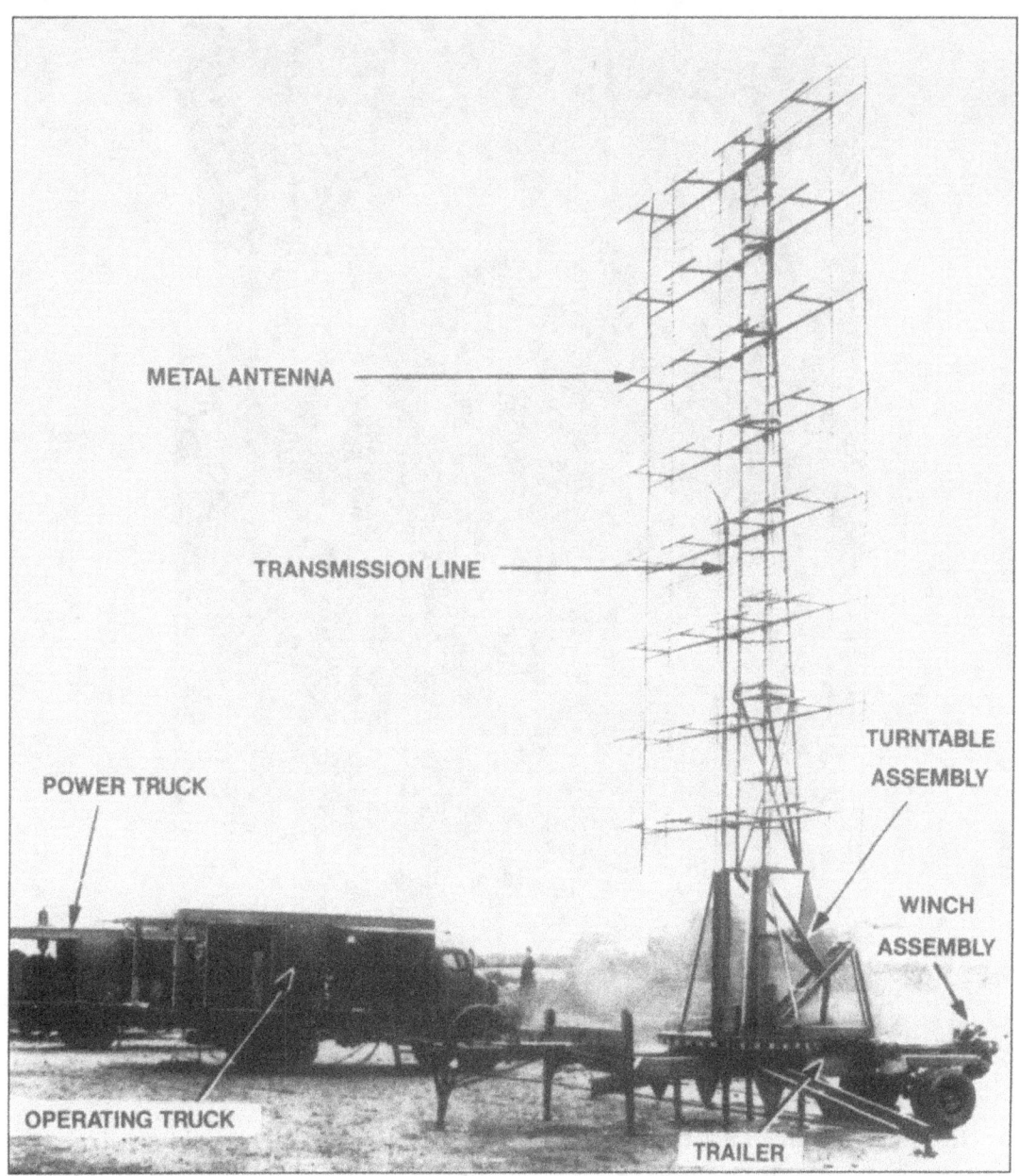

A mobile SCR-270 radar set. On December 7, 1941, one of these sets detected Japanese aircraft approaching Pearl Harbor. Unfortunately, the detection was misinterpreted and ignored. (Photograph courtesy Historical Electronics Museum, Inc.)

The SCR-270 operating position shows the antenna positioning controls, oscilloscope, and receiver. (Photograph courtesy Historical Electronics Museum, Inc.)

In addition to an air defense infrastructure, a scientific and industrial infrastructure also was developed. At the Massachusetts Institute of Technology (MIT), a cadre of scientists and engineers worked at what was called the "Radiation Laboratory" to develop and test new radar systems. During the war, the Radiation Laboratory produced some 150 distinct military radar systems that had land, sea, and air applications. Because of its contribution to the war effort, MIT was positioned to contribute solutions to America's air defense problems in the post-war era.[7]

Radars designed by MIT were not needed on the home front during World War II. American victory at Midway Island and an allied invasion into North Africa calmed

The Evolution of Air Defense (1918–1959)

An SCR-271 fixed radar site. (Photograph courtesy Historical Electronics Museum, Inc.)

Reports from radars and Ground Observer Corps lookouts were fed into information centers. This photograph shows the filter board of the New York Information Center. (Photograph courtesy Air Force History Support Office.)

public concerns about enemy bombings. By 1943, air defense on the home front had become a low priority. As a result, the ADC was disestablished.

Although domestic air defenses were never tested during the war, the Battle of Britain, coupled with Army Air Forces difficulties in penetrating German airspace, demonstrated that effective air defense could prove potent. However, the dropping of atomic bombs on Hiroshima and Nagasaki also restored confidence in the doctrine of offensive operations. At the conclusion of the war, all air defenses were shut down.[8]

The Post-War Era

During World War II, consideration had been given to how America's air defense system would operate during the post-war era. Subsequently, the Army Air Forces received

The Evolution of Air Defense (1918–1959)

George E. Stratemeyer when he was a Major General. Stratemeyer was the first post-WWII commander of the Air Defense Command. (Official U.S. Air Force photograph courtesy Air Force History Support Office.)

responsibility for manning, training, equipping, and deploying the fighter forces and needed warning radar stations. To organize its resources, the Army Air Forces reorganized to form the SAC, the TAC, and the reestablished ADC.[9]

Activated on March 27, 1946, the new ADC was directed to "organize and administer the integrated air defense system of the Continental United States."[10] With virtually no radars in operation and its fighter aircraft relegated to National Guard units, ADC's initial emphasis was on planning. Headquartered at Mitchel Field, New York, the new organization came under the command of Lieutenant General George E. Stratemeyer.

For air defense planners, the late 1940s proved a challenging period. Although the Cold War was becoming a reality, national leaders did not acknowledge the Soviet Union as an immediate military threat to the North American continent. American intelligence was aware that the Soviets were reproducing their own version of the American B-29 bomber; however, reconnaissance efforts in the late 1940s indicated that the Soviets were not constructing bomber bases in areas that could bring these planes within striking distance of the continental United States. Furthermore, the United States still retained sole possession of the atomic bomb. Air power projection advocates such as General Carl A. Spaatz and General Curtis LeMay viewed delivery of the atomic bomb as the primary Army Air Forces mission. They emphasized offensive air power as the best method of defense. Ground rules were set and would remain intact for years to come. Thus air defense planners competed against fellow Army Air Forces officers in the struggle to obtain appropriations.

For the ADC, the appropriations battle was a difficult struggle. In the immediate post-war era severe budget cuts rocked the entire military establishment. With reduced resources, Army Air Forces' Chief of Staff General Carl Spaatz provided support to SAC and TAC at the expense of ADC. Lieutenant General Earle E. Partridge, the Assistant Air Chief of Staff for Operations, felt that it would be a mistake for the Army Air Forces to give the public an impression, less than a year after Japan's surrender, that an air attack was anticipated. Instead, Partridge argued that funds should be used for research and development of longer-range radars.[11]

With the Soviet Union posing a potential threat, in 1946 Douglas Aircraft Company's Research and Development (RAND) Project (later to become known as the RAND Corporation) was asked by the Army Air Forces to appraise the air defense problem. While the RAND Project conducted its study, the Army Air Forces directed the ADC to draft a proposal to employ existing equipment. In October and November 1946, Lt. Gen. Stratemeyer submitted two proposals. The October proposal was a short-term plan that concentrated air defense forces in the northeast and northwest. The November proposal was a longer-term plan, calling for the use of twenty-four radars to be installed by 1949 to guard the approaches to five strategic areas that encompassed the northeast, the Chicago-Detroit area, and the three west coast cities of Seattle, San Francisco, and Los Angeles. However, in testimony before the House Appropriations Committee on March 6, 1947, General Spaatz suggested that the best defensive strategy was to attack the enemy bombers at their home airfields. With a defense reorganization pending that promised the creation of an independent United States Air Force, Spaatz advised the ADC commander, Lt. Gen. Stratemeyer, not to press demands. Still, planning continued and in April 1947, ADC proposed a network of 114 radars.[12]

In July 1947, the RAND Project issued a preliminary report recommending against earlier proposals that had called for immediate deployment of a radar net using World War II vintage equipment. Because an enemy air attack was considered highly improbable, the RAND Project recommended a minimal air defense. The military reorganization prompted by the National Security Act of 1947 froze any consideration for radar deployment during mid-1947 as the services settled down to redefine their missions. Finally, on November 12, 1947, Secretary of Defense James V. Forrestal announced that planning was underway for a national early warning radar network. Nine days later, General Spaatz approved the blueprint calling for a radar fence plan of 374 radar stations and 14 control centers to be built throughout the continental United States and an additional 37 stations and 4 control centers to be placed in Alaska. Called Project SUPREMACY, the plan predicted that with immediate funding, the system would be operational by mid-1953. Under this scheme, the radar stations would report intruding aircraft to the regional control center that in turn alerted interceptor aircraft. Once the interceptor aircraft were airborne, the radar stations would assume control and vector the interceptors against the attackers.[13]

The proposed Project SUPREMACY represented an enormous leap from what existed at the end of 1947. At that time ADC operated only two radar stations: one at Arlington, Washington, and one at Half Moon Bay near San Francisco, California. Manned by the 505th Aircraft Control and Warning (AC&W) Group, these stations worked with fighter squadrons to perfect ground-control and interception techniques. The experience gained from operating these two sites proved invaluable to air defense planners who were designing the nationwide system.[14]

The Evolution of Air Defense (1918–1959)

As Project SUPREMACY was undergoing consideration, relations with the Soviet Union continued to sour. In February 1948, there was a Communist coup in Czechoslovakia. In China, Communist forces continued to gain ground against Chiang Kai-shek. Air Force intelligence warned that the Soviets were preparing to conduct a surprise attack. On March 27, 1948, General Spaatz, concerned about the vulnerability of the Atomic Energy Commission plant at Hanford, Washington, ordered the recently placed ADC radars at Arlington, Spokane, Neah Bay, and Hanford, Washington, and at Portland, Oregon, to begin operating on a 24 hour-a-day basis. Due to insufficient personnel and materiel resources, round-the-clock operations in the northwest proved beyond ADC's capability. Despite these problems, ADC was ordered to take AN/CPS-5 and AN/TPS-1B/1D radar sets out of storage for operation in the northeast and in Albuquerque, New Mexico. By August, radars had been placed at Twin Lights and Palermo in New Jersey, and at Montauk, New York. In September 1948, the Air Force ordered thirteen additional World War II radars to be placed in operation over an area stretching from Maine to Michigan. Along with the previously sited radars, these sets became incorporated into what became known as the Lashup system. Lashup was an appropriate name for the system as World War II vintage radar antennas were literally lashed to the top of wooden

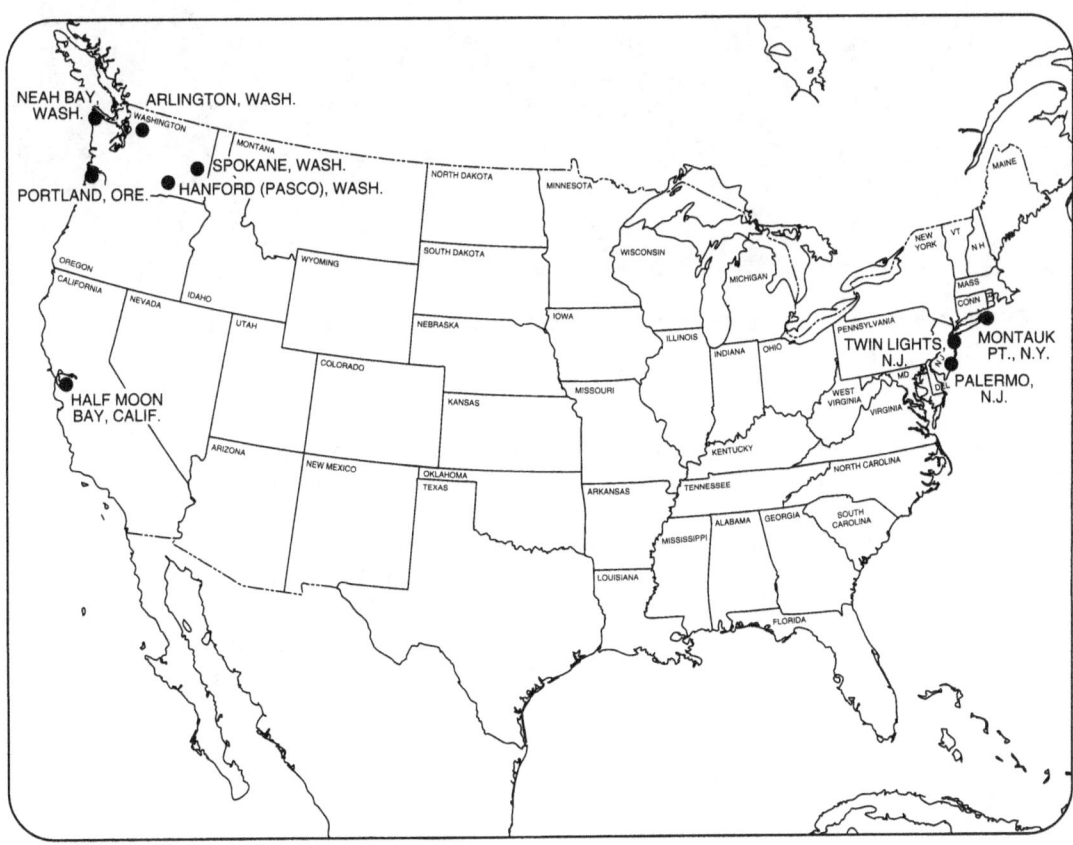

Deployment of Air Defense Radar, June 1948. (Map courtesy of Air Combat Command.)

AN/CPS-5 radar; the basic unit of the Lashup system. (Photograph courtesy Air Force History Support Office.)

platforms. In addition to the temporary antenna towers, Quonset huts and short-term wooden structures were built to house the equipment and radar operators.[15]

Despite further deterioration in the relationship between the United States and the Soviet Union, and the imminent fall of China, the price for Project SUPREMACY was still considered too costly by Defense Secretary Forrestal. The Project SUPREMACY plan ultimately was never implemented. To address the Secretary's financial concerns and still move ahead to deploy a radar network, in mid-1948 Air Force Vice Chief of Staff Muir S. Fairchild appointed air defense expert Major General Gordon P. Saville to a position as ADC Headquarters Special Projects Officer. After nearly two months, the team that Saville brought together produced a new plan that argued for deployment of a radar network as the first step in developing a credible air defense system.

Saville proposed a system of seventy-five radar stations and ten control stations in the continental United States and ten radar stations and a control center for Alaska. To

Saville, this plan represented a starting point. The seventy-five-station system eventually was dubbed the "permanent network." Additional stations could be built later to provide the coverage envisioned under Project SUPREMACY.

The plan found an advocate in Colonel Charles Lindbergh. Having been placed on active duty as a special consultant to evaluate technical and operational matters, Lindbergh argued that SAC bombers needed to fly against actual radar networks to provide a realistic assessment of their power projection capability. Thus a radar network would serve not only to detect intruders, but also to provide SAC a test bed to develop tactics to avoid detection. In October 1948, Forrestal approved Saville's plan.[16]

Forrestal's approval came at a time when President Harry S. Truman sought further cuts in the defense budget. The Air Force adjusted to the cuts by maintaining support for SAC and consolidating TAC and ADC into the Continental Air Command (CONAC). Formed on December 1, 1948, the new organization was headed by General Stratemeyer. Now a component command of CONAC, ADC was commanded by Maj. Gen. Saville.[17]

Major General Gordon P. Saville. Saville was a pioneer in air defense theory and would later command ADC. (Photograph courtesy Air Force History Support Office.)

Reflecting the President's interest in fiscal austerity, Louis Johnson, Forrestal's replacement as Secretary of Defense, kept the request for radar funding out of the supplemental 1949 and Fiscal Year (FY) 1950 budgets. Events in late 1949 began to arouse public concern over air defense. In August, under the encouragement of the Air Force, Boeing Company announced plans to shift B-47 production from Seattle to the less-vulnerable Wichita, Kansas. This announcement drew protest from Seattle leaders as well as Alaska's governor. It was their opinion that the Boeing move represented a tacit admission by the Air Force of the vulnerability of Alaska and the northwest to Soviet attack. During hearings before the House Armed Services Committee on the B-36 program, Navy leaders also questioned the Air Force's lack of expenditures on air defense. Finally, with President Truman's announcement on September 22, 1949, of a recent Soviet detonation of an atomic bomb, public interest in air defense became rampant.

Money was made available in the FY 1950 budget to start air defense construction. In addition, Congress granted the Air Force authority to transfer money from other projects to expedite building the permanent network.[18]

On December 2, 1949, the Air Force directed the Army Corps of Engineers to proceed with construction of the first twenty-four radar sites on Saville's seventy-five-site list. Areas covered by these sites included northeastern, midwestern, and western metropolitan regions, and Atomic Energy Commission sites in Washington and New Mexico. Many of these locations already had temporary radars operating as part of the Lashup system. By mid-1950, forty-four Lashup installations already were operating around the strategically important areas. Once the permanent network stations became operational, the Lashup stations would be retired.[19]

Also in 1950, other steps were taken to improve the nation's air warning capabilities. The Ground Observer Corps (GOC) was reestablished. In addition, Canada and the United States agreed to extend the American radar network into Canada. To complete this effort, the United States cooperated in constructing, equipping, and operating some of these

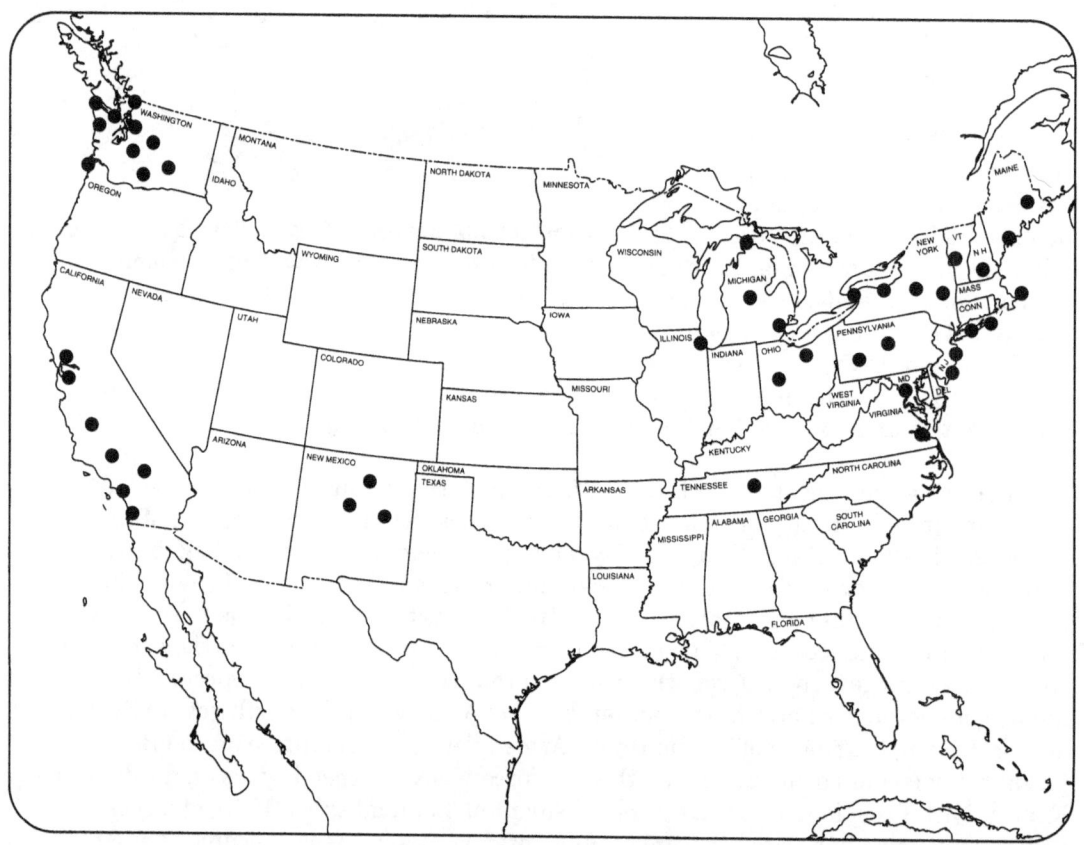

Deployment of Air Defense Radar, April 1950 (Lashup). (Map courtesy of Air Combat Command.)

The Evolution of Air Defense (1918–1959)

THE GROUND OBSERVER CORPS

The Ground Observer Corps (GOC) traced its roots to World War II when 1.5 million civilian volunteers were enrolled by the Army Air Forces to man 14,000 observation posts positioned along the nation's coasts. With limited radar detection capability, the GOC's mission was to visually search the skies for enemy aircraft attempting to penetrate American airspace. With the declining threat to America from German and Japanese air forces, the Army Air Forces disestablished the GOC in 1944.

In February 1950, Continental Air Command Commander General Ennis C. Whitehead proposed the formation of a 160,000 civilian volunteer GOC to operate 8,000 observation posts scattered in gaps between the proposed radar network sites. With the belief that the Korean War served as a precursor to a possible Soviet attack, ADC had little difficulty recruiting volunteers. In 1951, some 210,000 GOC volunteers manning 8,000 observation posts and twenty-six filter centers were tested for the first time in nationwide exercises. The time recorded for a sighting report to reach the Ground Control Interception centers through the filter centers in this and subsequent drills was unimpressive. Subsequently, the scope of Whitehead's plan was expanded to recruit more volunteers to man more observation posts on a continuing basis. This revised GOC plan, dubbed "Operation SKYWATCH," was initiated on July 14, 1952. Eventually over 300,000 volunteers stood alternating shifts at 16,000 observation posts and seventy-three filter centers. The Air Force used a variety of means to recruit volunteers, including radio. One radio spot announced:

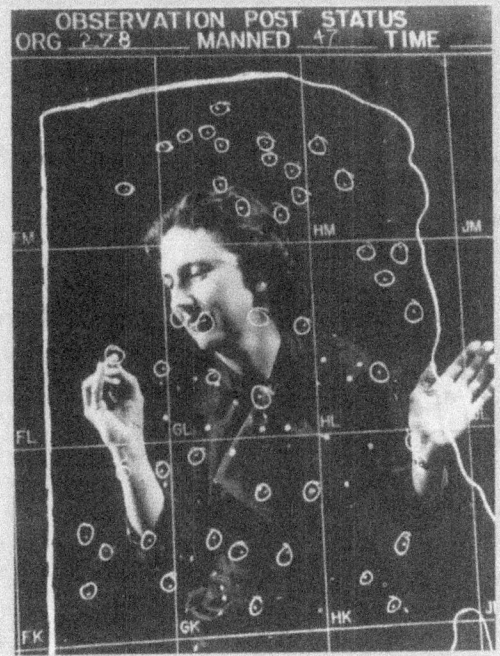

This photograph, taken in 1956, shows a Ground Observer Corps volunteer on duty in one of the seventy-three Continental Air Defense Command Filter Centers plotting the track of an aircraft using reports transmitted to her by other volunteers manning observer posts within the area. When the track was established, the information was forwarded to the Air Defense Direction Center, where the decision to scramble interceptor aircraft for identification purposes was made. (Official U.S. Air Force photograph courtesy Air Force History Support Office.)

> "It may not be a very cheerful thought but the Reds right now have about a thousand bombers that are quite capable of destroying at least 89 American cities in one raid.... Won't you help protect your country, your town, your children? Call your local Civil Defense office and join the Ground Observer Corps today."

Source: Schaffel, *Emerging Shield*, pp. 158–159.

stations on the northern side of the U.S.-Canadian border as well as those on the southern side. This string of stations straddling the border became known as the "Pinetree Line."[20]

By the late 1950s, deployment of the short-range AN/FPS-14 radar resolved the problem of detecting low-flying planes. Dozens of AN/FPS-14s and the follow-on model AN/FPS-18s were deployed at sites between the long-range permanent and mobile radar stations. As a result of this technological improvement, the GOC was deactivated on January 31, 1959.

Building the Network

On June 25, 1950, North Korea launched an invasion of South Korea, drawing the United States into a war that would last for three years. Believing that the North Korean attack could represent the first phase of a Soviet-inspired general war, the Joint Chiefs of Staff ordered Air Force air defense forces to a special alert status. In the process of placing forces on heightened alert, the Air Force uncovered major weaknesses in the coordination of defensive units to defend the nation's airspace. As a result, an air defense command and control structure began to develop and Air Defense Identification Zones (ADIZ) were staked out along the nation's frontiers. With the establishment of ADIZ, unidentified aircraft approaching North American airspace would be interrogated by radio. If the radio interrogation failed to identify the aircraft, the Air Force launched interceptor aircraft to identify the intruder visually. In addition, the Air Force received Army cooperation. The commander of the Army's Antiaircraft Artillery Command allowed the Air Force to take operational control of the gun batteries as part of a coordinated defense in the event of attack.[21]

On July 11, 1950, the Secretary of the Air Force requested approval from the Secretary of Defense to expedite construction of the second segment of twenty-eight stations for the permanent network. Most of these stations provided additional coverage to eastern, midwestern, and western regions of the country. Receiving the Defense Secretary's approval on July 21, the Air Force directed the Corps of Engineers to proceed with construction.

The remaining twenty-three permanent network sites were approved for construction later in 1950. Located primarily in Minnesota, North Dakota, and Montana, these sites formed the American component of the Pinetree Line. In September 1950, Congress provided a supplemental appropriation of $40 million to fund construction and equip the sites with the newest radars.[22]

Before a closed session of the House Armed Services Committee on July 27, 1950, Continental Air Command Vice Commander General Charles T. Myers pledged that the seventy-five stations would be finished by July 1, 1951. This promise proved impossible to keep. Lack of coordination between various Air Force commands and the Army Corps of Engineers, funding problems, manpower shortages, building material and spare part shortages, as well as a strike at General Electric's radar fabrication plant all slowed progress. By the end of December 1950, the completion date for the permanent network had been set back six months.[23]

The Evolution of Air Defense (1918–1959)

> ## Construction Challenges
>
> The Army Corps of Engineers had to overcome several difficulties in positioning many of the radar stations on remote, inaccessible mountain peaks. Construction of the site at Cape Newenham, Alaska, started in 1950, provides an example of some of the more extreme challenges faced by the Corps. Located on a small peninsula on the southwest coast of Alaska, Cape Newenham could only be reached by air and by sea. The site was divided into lower and upper camps. Because the terrain made road construction difficult, a tramway had to be built to move men and materials up the mountain. Construction of Cape Newenham was finally completed in 1954.
>
> Much work still needed to be done once the construction crews left a site. Installation and calibration of the radars often took longer than anticipated. For example, the Air Force occupied the first permanent site at McChord Air Force Base in Washington in the fall of 1950. However, the radar installation was not completed until February 1951, and calibration, training, and integrating the site into the air defense network took another six months.
>
> Delays also plagued mobile radar deployment. Problems included site placement indecision and a variety of construction problems. For example, site SM-149 at Baker, Oregon, experienced a change of plans when engineers discovered it would cost $100,000 to run a water pipeline up the mountain. Consequently, the planned cantonment area was moved off the mountaintop. At site SM-152 at Geiger Field, Washington, engineers had not accounted for winds up to ninety miles per hour and snow that could reach twenty feet deep. Plans were redrafted.
>
> ---
>
> Sources: D. Colt Denfield, *The Cold War in Alaska: A Management Plan for Cultural Resources*, (Alaska: U.S. Army Corps of Engineers Alaska District, 1994), pp. 151–152; McMullen, *Radar Programs*, pp. 36–37, 62, 65–66.

As construction of the permanent network proceeded, Congressional concerns about air defense prompted a reorganization of the Air Force. On January 1, 1951, the Air Force reestablished ADC as a major command to be headquartered at Ent Air Force Base (AFB) in Colorado.

In the wake of Communist China's intervention in Korea, Congress approved President Truman's request for supplemental funds that included appropriations for a mobile radar network to supplement the permanent network.[24] In July 1951, ADC received approval to install forty-four mobile radars to provide protection for key SAC bases. ADC planned to have these radars operating the following July. The permanent sites were designated P-sites and the mobile sites were designated M-sites. In January 1952, ADC decided to position some of the mobile radars in conjunction with permanent sites to form a double perimeter around strategic areas. The moniker "mobile site" was a misnomer. Understanding that these mobile sites would remain in the same location for the long term, ADC directed the Army Corps of Engineers to build permanent support facilities at

Searching the Skies: The Legacy of the United States Cold War Defense Radar Program

Finley Air Force Station, North Dakota, became operational in 1952 and typifies a permanent network site. (Official U.S. Air Force photograph.)

each site. As with the permanent network, mobile radar deployment was slowed due to procurement problems.[25]

As nearly all stations of the permanent network reached operational status, the Air Force approved the second phase of the mobile radar program on October 18, 1952. These stations were designated as second mobile or SM-sites. Meanwhile, the Navy was asked to provide radar picket ships to cover coastal approaches and the Air Force began to purchase EC-121 Lockheed Constellation planes to provide additional radar coverage.[26]

The Debate

Even with a fully capable radar system serving as the foundation of an air defense infrastructure, the Air Force claimed the United States could stop only thirty percent of an attack, at best. However, in 1950, Dr. George E. Valley, Jr., an MIT physics professor and member of the U.S. Air Force Scientific Advisory board, led a committee that more realistically concluded that the air defense system could stop only about ten percent of an attack. The Valley Committee recommended solutions that included establishing an air defense laboratory at MIT. This laboratory would employ new technologies to improve this percentage rate. The Air Force expressed interest in establishing such a laboratory. However, resistance existed at MIT by faculty who objected to the university's continuing

The Evolution of Air Defense (1918–1959)

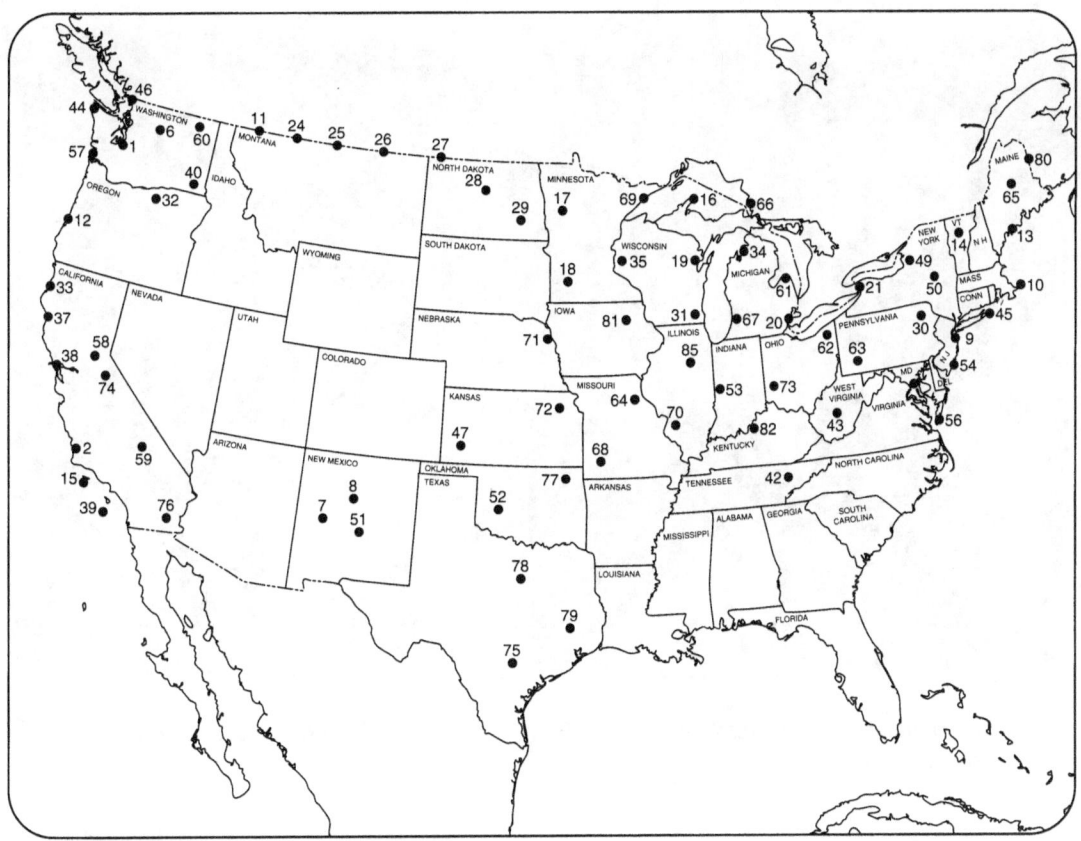

Progress of the Permanent System, June 1952. (Map courtesy of Air Combat Command.)

support of military research and development. MIT President James R. Killian convened a study group, named "Project CHARLES," to examine the laboratory proposal. In addition to agreeing that MIT should host an air defense laboratory, Project CHARLES concluded that technology existed that was capable of surmounting the air defense problem. By establishing a laboratory dedicated to air defense, MIT took on a project with a budget twice that of its undergraduate teaching program. The air defense laboratory at MIT eventually became known as the Lincoln Laboratory.[27]

In partnership with the Air Force, Cambridge Research Laboratory, and IBM, the Lincoln Laboratory immediately began work to modify a Whirlwind computer that was being developed for the Navy's use in performing air defense command and control functions. What emerged was the AN/FSQ-7, otherwise known as Whirlwind II. In 1951, Whirlwind II was first tested by placing the computer at a control center in Cambridge to receive data from a long-range and several short-range radars set up on Cape Cod. Tests proved promising, but years of development still lay ahead. The key breakthrough was the development of magnetic-core memory that vastly improved the computer's reliability. When the previous electrostatic-storage-tube memory was replaced

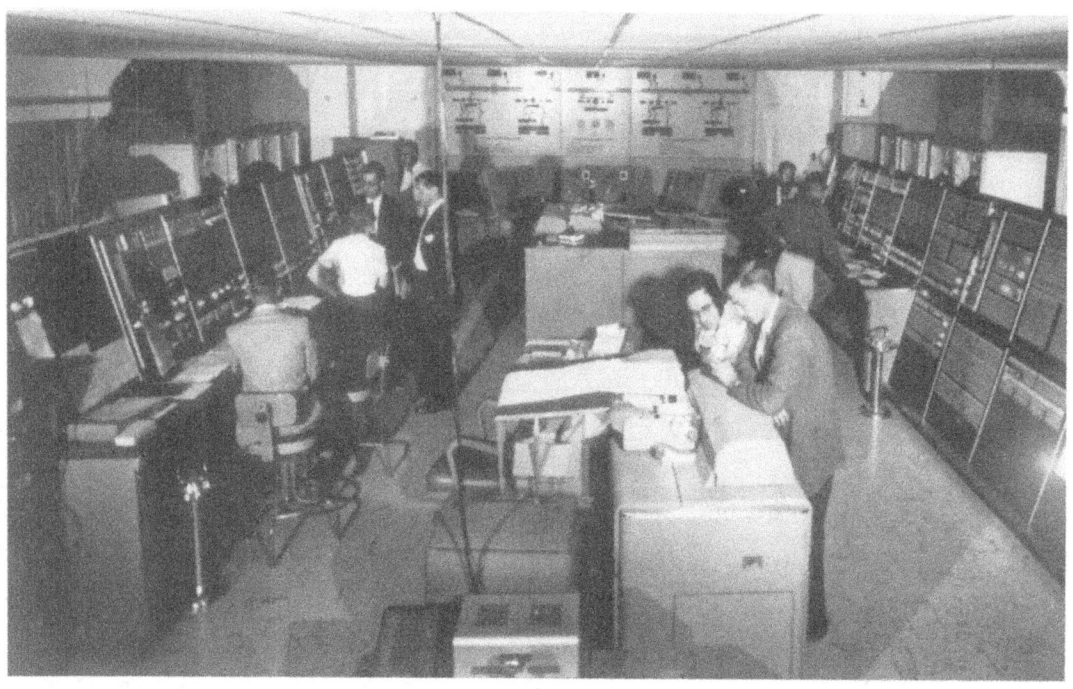

The Maintenance Control Console of the AN/FSQ-7 Computer. (Photograph courtesy Air Force History Support Office.)

by magnetic-core memory, operating speed doubled and input speed quadrupled. More significantly, maintenance time for the core memory dropped from four hours per day to two hours per week.[28]

On April 16, 1952, after receiving reports from Alaska and Maine of unidentified incoming aircraft, ADC Headquarters issued an air defense readiness alert that caused hundreds of pilots to scramble to their planes and guncrews to man their antiaircraft guns. The threat was later determined to be false. Air defense planners were forced to acknowledge limited capability to evaluate threats and respond. Telephone and teletype communications were too slow to keep an air defense commander cognizant of an evolving air battle. In the wake of the false alert, defense planners decided to reevaluate the emerging air defense system.[29]

A "Summer Study Group" met from June through August 1952 at MIT. Twenty scientists and engineers, along with several consultants, considered current and future threats, such as Intercontinental Ballistic Missiles (ICBMs), to the United States. Regarding the current threat, the group concluded that early warning was critical for a successful defense. They recommended the establishment of a Distant Early Warning (DEW) Line across the northern tier of the North American continent. They further concluded that automation of command and control through the introduction of computers, such as Whirlwind II, would give air defense commanders valuable minutes to properly deploy interceptor aircraft.[30]

The Evolution of Air Defense (1918–1959)

The patch panel between the central computer and the display consoles.
(Photograph courtesy Air Force History Support Office.)

Throughout late 1952, Air Force officials and scientists vigorously debated the DEW Line proposal. Opponents feared that the United States would build a Maginot Line[a] at the expense of SAC. The debate was internal. Neither Congress nor the American people were aware of the proposals being discussed. At the end of 1952, President Truman stepped in and signed the National Security Council (NSC) directive 139. NSC 139 directed the construction of the DEW Line. After reevaluating Soviet atomic bomb and bomber production rates, NSC 139 preparers identified 1955 as a period of maximum danger. Facing this imminent Soviet threat, defense planners considered an effective air defense warning system to be essential.[31]

However, newly elected President Dwight D. Eisenhower desired a reduction in defense spending and a change of priorities. The new administration no longer considered 1955 to be a period of maximum danger. Air defense again was scrutinized by a committee headed by Bell Telephone Laboratories president Mervin J. Kelly. While the Kelly Committee reviewed Summer Study Group recommendations, the American people became aware of the debate through congressional testimony and press coverage. On March 6,

---

[a] The term Maginot Line had deep symbolism in this era. The French had built an extensive fortification line facing the German border that was rendered useless by a German flanking attack through Belgium in the Spring of 1940. Analysts have argued that a French investment in a mobile offensive force rather than a static defense could have prevented defeat.

1953, Air Force Chief of Staff General Hoyt S. Vandenberg testified before Congress against funding for defensive systems at the expense of improving American air offensive capabilities. Newspaper columnists Joseph and Stewart Alsop strongly disagreed. In their *New York Herald Tribune* columns, the Alsop brothers published accounts of the Summer Study Group recommendations and the subsequent deliberations portraying the Air Force, and specifically SAC, as villains suppressing technological advances. In May 1953, the Kelly Committee issued a report that seemed to vindicate both sides; both sides interpreted the report in their favor.

Not pleased with the Kelly Committee findings, Defense Secretary Charles E. Wilson appointed retired Army Lieutenant General Harold R. Bull to lead another committee. In July 1953, the Bull Committee submitted a report that supported many of the Air Force planning efforts. The report recommended construction of a sensor line across mid-Canada as a top priority. Second priority would go to building the DEW Line (if experimental tests in Alaska proved it workable); deploying an automated command and control system; building an unmanned network of short-range, low-altitude, gap-filler

The Evolving Air Defense Organization: CONAD

As ADC modernized and expanded its radar network, the command structure ADC operated within underwent constant change. As a result of a 1948 Joint Chiefs conference held in Key West, Florida, each military branch was assigned separate air defense responsibilities. The Air Force received primary responsibility for continental air defense. The Army contributed antiaircraft artillery. The Navy provided radar picket ships. Two years later, an agreement between the Army and Air Force Chiefs of Staff placed Army antiaircraft units under Air Force operational control to ensure coordination. Satisfied with this arrangement, the Air Force resisted any ideas of forming a multi-service or joint command to oversee air defense.

However, Chairman of the Joint Chiefs of Staff, Admiral Arthur Radford, firmly believed in the necessity of a joint command to coordinate the nation's resources against a possible Soviet attack. In discussing this concept with Air Force Chief of Staff General Nathan Twining, Radford foresaw a joint organization containing components from the three military services commanded by an Air Force general. Impressed, Twining reversed the Air Force's previous position and directed his chief of staff to begin drafting this new command's charter and directives. Eventually, ADC performed much of the staff work to design the new organization.

Although the Army protested the creation of a joint organization that had an Air Force general as permanent head, the Joint Chiefs accepted the Air Force's proposed organizational design and directed the establishment of the Continental Air Defense Command (CONAD). Headquartered in Colorado Springs, Colorado, this organization was activated on September 1, 1954.

The Evolution of Air Defense (1918–1959)

radars to replace the Ground Observer Corps; and implementing an improved aircraft identification system. Due to questioning by the Joint Chiefs regarding the priorities of the Bull Report, the National Security Council postponed consideration of the findings until September 1, 1953. In the wake of the Soviet explosion of a hydrogen bomb in August 1953, the National Security Council approved an amended version of the Bull Report. The approval document, NSC 159/4, proved significant; the Air Force received support to proceed with the development of an air defense structure.[32]

Thus the path was cleared for ADC to request funding for a third phase of mobile radars that would prevent an end-run around the northern-oriented detection belt. These radars, to be placed along the east and west coasts and in the south, were called third mobile or TM-sites. The Air Force approved this request and budgeted for the integration of an automated command and control system into the air defense network.[33] This system was being developed at Lincoln Laboratory.

The emphasis on defense approved in NSC 159/4 seemed to contrast with the policy promulgated in late 1953 in NSC 162, a policy that became known as the "New Look." The New Look had an emphasis on massive retaliation. Yet, the two policies were complementary. Eisenhower recognized that if America was to deter war through massive retaliation, it needed air defenses to ensure the survival of its retaliation force. Thus strong air defense contributed to the credibility of the American strategy.[34]

Planning went ahead for a future defensive structure. However, execution of current plans lagged due to construction and equipment procurement problems. As of late 1953, not one mobile radar station was operational and sites were still being surveyed for the second phase of mobile radars.[35]

Improving Command and Control

The permanent network depended on each radar site to perform GCI functions or pass information to a nearby GCI center. For example, information gathered by North Truro Air Force Station on Cape Cod was transmitted via three dedicated land lines to the GCI center at Otis AFB, Massachusetts, and then on to the ADC Headquarters at Ent AFB, Colorado.

The facility at Otis AFB was a regional information clearinghouse that integrated the data from North Truro and other regional radar stations, Navy picket ships, and the all-volunteer GOC. The clearinghouse operation was labor intensive. The data had to be manually copied onto Plexiglas™ plotting boards. The ground controllers used this data to direct defensive fighters to their targets. It was a slow and cumbersome process, fraught with difficulties. Engagement information was passed on to command headquarters by telephone and teletype.

At Ent AFB, the information received from the regional clearinghouses was then passed on to enlisted airmen standing on scaffolds behind the world's largest Plexiglas™ board. Using grease pencils, these airmen etched the progress of enemy bombers onto the

Searching the Skies: The Legacy of the United States Cold War Defense Radar Program

(1)

(2)

Potentially hostile aircraft were first detected and the range tracked using a search radar (1). The altitude of the intruder was then determined using a height-finding radar (2). This information was passed from the radar station (3) to the regional command center (4). The regional command center passed the information on to CONAD, later NORAD, at Ent Air Force Base Colorado (5). If the radar contact was deemed a potential threat, the regional command center would direct interceptor aircraft to scramble (6). Radar station personnel tracking the potentially hostile aircraft would then vector the interceptor aircraft to the target. (Official Air Force photographs courtesy USAF Museum [1–3] and National Archives [4–6].)

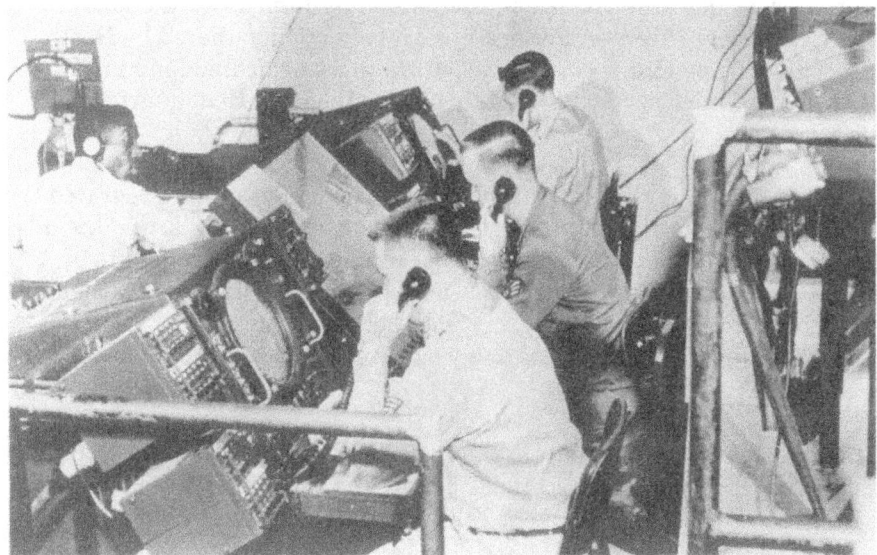
(3)

The Evolution of Air Defense (1918–1959)

(4)

(6)

(5)

back of the Plexiglas™ board so that air defense commanders could evaluate and respond. This arrangement impeded rapid response to the air battle.[36]

At the Lincoln Laboratory development continued on an automated command and control system centered around the 250-ton Whirlwind II (AN/FSQ-7) computer. Containing some 49,000 vacuum tubes, the Whirlwind II became a central component of the SAGE system. SAGE, a system of analog computer-equipped direction centers, processed information from ground radars, picket ships, early-warning aircraft, and ground observers onto a generated radarscope to create a composite picture of the emerging air battle. Gone were the Plexiglas™ boards and teletype reports. Having an instantaneous view of the air picture over North America, defense commanders would be able to quickly evaluate the threats and effectively deploy interceptors and missiles to meet the threat.

By 1954, with several more radars in the northeast providing data, the Cambridge control center (a prototype SAGE center) gained experience in directing F-86D interceptors against B-47 bombers performing mock raids. Still much development, research, and testing lay ahead. Bringing together long-range radar, communications, microwave electronics, and digital computer technologies required the largest research and development effort since the Manhattan Project. During its first ten years, the government spent $8 billion to develop and deploy SAGE. By 1958, Lincoln Laboratory had a professional staff of 720 with an annual budget of $22.5 million, to conduct SAGE-related work. The contract with IBM to build sixty production models of the Whirlwind II at $30 million each

---

THE EVOLVING AIR DEFENSE ORGANIZATION: NORAD

As early as 1951, Royal Canadian Air Force officers served liaison duty with the U.S. Air Force Air Defense Command (ADC) at Colorado Springs. However, the concept of Canada integrating its air defense command with the United States posed difficult problems for Canadian leaders concerned about sovereignty. Beginning in 1954, study groups on both sides of the border examined the issue and recommended a combined air defense command structure to the military and political leadership. On August 1, 1957, Canada and the United States issued a joint communique and in September established the North American Air Defense Command (NORAD) at Ent AFB in Colorado Springs. Component commands of NORAD included the U.S. Army ADC, U.S. Naval Forces CONAD, the Air Force ADC, and the Air Defense Command of Canada. CONAD remained in existence to handle U.S. responsibilities outside of NORAD jurisdiction.

Sources: Kenneth Schaffel, *The Emerging Shield: the Air Force and the Evolution of Continental Air Defense, 1845–1960*, pp. 243–245, 253; *A Hand Book of Aerospace Defense Organization, 1946–1986*, p. 5; Lawrence J. Kilbourne and Thomas Fuller, "An Important Anniversary: Thirty Years of Successful U.S.-Canadian Partnership in NORAD," *Canadian Defense Quarterly*, (Summer 1987), pp. 36–40.

provided about half of the corporation's revenues for the 1950s and exposed the corporation to technologies that it would use in the 1960s to dominate the computer industry. In the meantime, scientists and electronic engineers in the defense industry strove to install better radars and make these radars invulnerable to electronic countermeasures (ECM), commonly called jamming.[37]

Improving the Radar Network

In addition to SAGE center development, progress continued on mobile and other radar network installations. On December 6, 1954, Site M-129 at MacDill AFB, Florida, became the first mobile radar site to achieve operational status.

By the end of 1955, thirteen Phase I mobile (M-site) and one Phase II second mobile (SM-site) stations were operational. They joined seventy-five stations of the permanent network. Along with these stations, other radar lines were constructed in Canada. The Pinetree Line, operational in 1954, straddled the United States/Canadian border and consisted of over thirty stations. The United States paid two-thirds of the costs and provided most of the manpower. North of the Pinetree Line was the Mid-Canada Line, built by the Canadian government. The Mid-Canada Line consisted of an unmanned microwave fence designed to detect flyovers.

The DEW Line began with an experimental station at Barter Island, Alaska, in early 1953. During the summer, work began on an eighteen-site test line across northern Alaska and northwestern Canada. By 1954, successful tests at these stations spurred extending the line across the Canadian arctic. By the end of 1957, fifty-seven stations were completed in a very costly and challenging construction effort.[38]

As the radar network expanded to the top of the North American continent, defensive planners expressed concern that the radars in operation would not be capable of detecting new high-altitude aircraft. The AN/FPS-3, a radar designed and built after World War II, could detect targets only up to 55,000 feet. In the early 1950s, technicians combined a device featuring a klystron tube with the radar to improve height-detection capability. With this device, called the GPA-27, the redesignated AN/FPS-3A radar could detect targets at 65,000 feet. Beginning in 1956, GPA-27 kits were installed at AN/FPS-3 sites. In addition to the older radars retrofitted with the GPA-27, new sets received the device as integral equipment. Designated as the AN/FPS-20, these new radars began to perform air search duties in 1957 and continued in service through the end of the Cold War.[39]

Another post-World War II radar, the AN/CPS-6, also faced obsolescence. Rather than invest much money to slightly improve the sets' performance, the Air Force decided to replace them with the AN/FPS-7 that was being developed for the Navy by General Electric (GE). The AN/FPS-7 held promise for detecting targets up to 100,000 feet. In 1955, ADC received authorization to acquire thirty-three of these sets. Development problems delayed deployment. The first set began operating at Highlands, New Jersey, in 1959.[40]

Engineers used modular construction for the stations along the 3,000-mile long DEW line. These photographs were taken in Canada in 1956. (Official U.S. Air Force photographs courtesy Air Force History Support Office.)

The Evolution of Air Defense (1918–1959)

An AN/FPS-20 radar set, in the foreground, located at Verona, New York, Test Site, Rome Air Development Center. This radar went into production in 1956 and was equipped with a klystron transmitter for dual-channel operations. (Official U.S. Air Force photograph courtesy Rome Laboratory History Office.)

In 1955, an inter-service study group named "Project LAMPLIGHT" reported that ECM could easily blind the current radar system. The study's conclusions were confirmed a year later when SAC bombers with ECM equipment blinded the ADC radar network during a mock attack. The Air Research and Development Command accepted the LAMPLIGHT Report and began developing a frequency-diversity (FD) radar. By giving the radar operator the ability to change the frequency of the radio wave emitted from the radar antenna, scientists and electrical engineers believed they could counter enemy jamming attempts. As with SAGE, years of research, development, and testing lay ahead.[41]

During the late 1950s another area of progress was the development and deployment of AN/FPS-14 and AN/FPS-18 gap-filler radars. Having a range of around sixty-five miles, these radars were placed in areas where it was thought enemy aircraft could fly low to avoid detection by the longer-range radars of the permanent and mobile radar

> ### DoD-FAA Cooperation
>
> A cooperative effort within the United States began in 1952 with talks between the Civil Air Administration (CAA) and ADC after a series of mid-air collisions involving civilian aircraft. A survey in 1956 revealed that the CAA planned to construct radars that overlapped areas already covered by ADC radars. Consequently, a Joint Radar Planning Group (JRPG) was formed to coordinate the selection of sites and radars for joint use to avoid resource duplication. JRPG agreed that AN/FPS-20 radars could support air traffic control in New York, Kansas City, Houston, Spokane, and possibly other areas. CAA ARSR-1 radars could provide ADC radar coverage around Denver, Salt Lake City, Atlanta, and possibly Miami and Oakland. This arrangement was formalized in an agreement between the Department of Defense and the Department of Commerce signed on January 9, 1958. By the end of 1959, ten ADC radars also performed air traffic control functions, and a Federal Aviation Administration (FAA) ARSR-1 radar at Richmond Naval Air Station (NAS), Florida, fed information into the air defense system.
>
> Source: McMullen, *Radar Programs*, pp. 116–117, 132–133.

networks. Gap-filler radar deployment peaked in December 1960 at 131 sites throughout the continental United States. Because the introduction of gap-filler radars alleviated the need for civilians to scan the skies for enemy bombers, the ADC disestablished the Ground Observer Corps on January 31, 1959 (as noted on page 22).[42]

Chapter 2

The Evolution to Aerospace Defense (1959–1979)

In August 1957, the Soviets successfully launched the SS-6 Sapwood ICBM. With an estimated range of 6,000 miles, the SS-6 represented a quantum leap in the Soviet rocketry program. American leaders were concerned as the Soviets now potentially had the capability to circumvent the North American air defenses. This concern was increased on October 4, 1957, when the Soviet Union launched Sputnik. With the advent of Sputnik, the general public became aware of America's potential vulnerability and placed U.S. defense programs in the spotlight.[43]

With this new threat on the horizon, work continued to improve the nation's ability to defend against a bomber attack. The first SAGE center became operational at McGuire AFB, New Jersey in June 1958. Air defenses reached a zenith in 1962. Although there were fewer combat and direction centers than originally planned, the air defense system conceived in the early 1950s was largely in place. By 1962 the SAGE system was completed. From the permanent and mobile radar construction programs, 142 primary radar stations and 96 gap-filler radar sites were operational in the United States and Canada providing data to the SAGE centers. Many of the primary radar stations hosted FD radars. The DEW Line across the northern continent was complete.

The SAGE combat and direction centers commanded a vast array of weapon systems. Forty-one interceptor squadrons numbering 800 aircraft, seven BOMARC missile squadrons, and scores of Army Nike missile battalions stood ready.[44]

In retrospect, it is easy to understand reasons for the decline of America's air defenses in the years following 1962. Technical advances threatened to make America's air defenses irrelevant. Speaking before the House Subcommittee on Department of Defense Appropriations in February 1966, Defense Secretary Robert McNamara stated:

> ... [T]he elaborate defenses which we erected against the Soviet's bomber threat during the 1960s no longer retain their original importance. Today, with no defense against the major threat, Soviet ICBM's, our anti-bomber defenses alone would contribute very little to our damage limiting objective and their residual effectiveness after a major ICBM attack is highly problematical. For this reason we have been engaging in the past five years in a major restructuring of our defenses.[45]

The introduction of ICBMs gave defensive planners a challenge they could not overcome, even though the Army spent billions of dollars to field an Antiballistic Missile

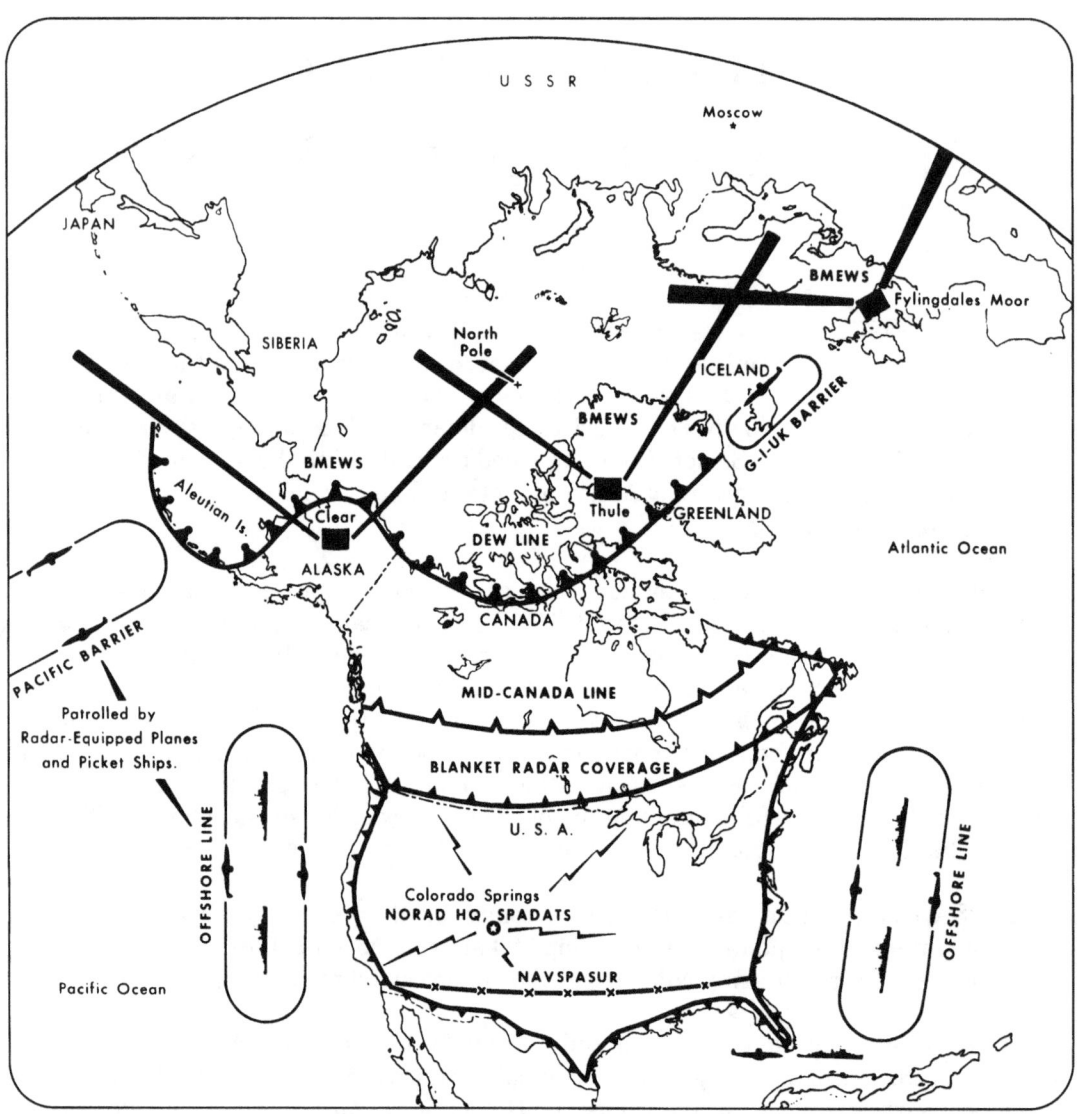

Map depicting United States defense sensors circa 1962. (Map courtesy USAF Museum.)

(ABM) system. Despite the Army program, Secretary McNamara felt that an attack could only be deterred through the assured destruction of any attacker. Consequently, billions of dollars were spent to upgrade strategic forces; SAC sunk 1,000 Minuteman silos into the western countryside and the Navy commissioned forty-one ballistic missile submarines. The Soviets countered with their own deployment of extensive rocket forces and missile-equipped submarines. Warning networks were upgraded only to allow strategic forces additional time to launch a retaliatory blow should the Soviets launch an attack. For the time being, the proponents of offense as the best defense had won their case.[46]

The Evolution to Aerospace Defense (1959–1979)

The Radar Network After Sputnik

The new missile threat did not remove the bomber threat. However, many in Congress felt that funds spent on bomber defenses were wasted. During 1959, funds for additional FD and gap-filler radars were cut. Cuts were also made to the SAGE command and control program.[47]

The funding prospects for air defense planners continued to look gloomy in 1960 as the Air Force advised NORAD that funding cuts expected in FY 1961 would force a revision of plans. Subsequently, NORAD identified twenty-six stations that could be released from the radar network. Some of the sites could be transferred to the FAA. The Air Force approved the plan and the deletions were made.[48]

Meanwhile, funding cuts and technical difficulties plagued the frequency-diversity radar program. Testing of the AN/FPS-24, AN/FPS-27, and AN/FPS-35 revealed serious design deficiencies. Technicians worried that the FD radars might not be compatible with the new SAGE system. Concerns were expressed that the high-power output would interfere with other electronic systems. These concerns were confirmed when passing radar beams of an AN/FPS-35 being tested at Montauk, Long Island, interfered with radio receivers and scrambled television signals over a six-mile radius. At Almaden, California, testing of an AN/FPS-24 radar could only be conducted at times when the local television stations were not broadcasting. As more FD radars began testing, complaints from television and radio station owners, as well as viewers and listeners, began to mount in Congress. To review the problem, the Air Force called for a two-day conference at Hanscom Field, Massachusetts, at the beginning of August 1962. After reviewing the problem, electronics experts concluded that the interference problems could be resolved if broadcasters followed a few simple procedures. Consequently, the path was opened for round-the-clock FD radar operations.[49]

Secretary of Defense Robert S. McNamara. (Photograph courtesy of Air Force History Support Office.)

The AN/FPS-24 used frequency-diversity techniques to counter electronic countermeasures. This radar became a component of the Semi-Automatic Ground Environment (SAGE) System. (Official U.S. Air Force photograph, courtesy Rome Laboratory History Office.)

While the ADC struggled to field FD radars, improvements to the existing network of AN/FPS-20 radars made it less susceptible to jamming. With the installation of Bendix-produced AN/GPA-102 or AN/GPA-103 kits, the performance of the radars improved to such a degree that they warranted redesignation. By the end of 1962, over one-third of the 131 primary radars within the air defense network were FD types or were electronic counter-countermeasure (ECCM)-modified AN/FPS-20s. Fifty radars were AN/FPS-7 sets that also had an ECCM capability.[50]

Radar Redesignations

Original System		Upgrade		New System
AN/FPS-20	+	AN/GPA-102	=	AN/FPS-64
AN/FPS-20	+	AN/GPA-103	=	AN/FPS-65
AN/FPS-20A	+	AN/GPA-102	=	AN/FPS-66
AN/FPS-20A	+	AN/GPA-103	=	AN/FPS-67

With few exceptions, by the end of 1962 air defense network radars provided data feeds into the completed SAGE command and control network. A national network, SAGE included eight regional combat centers and twenty-two direction centers scattered around the nation. SAGE designers built redundancy into the system, which gave each combat center the capability to coordinate defense for the whole nation. Meanwhile, direction centers evaluated data feeds from sector radar sites and directed aircraft and missiles against the threat. The final SAGE direction center became operational at Sioux City, Iowa, in December 1961.[51] SAGE centers allowed for a dramatic reduction of manpower over individual radar stations that once handled GCI functions. Manning levels dropped from nearly 200 to just over 100 men. Units designated as Aircraft Control and Warning Squadrons were renamed as Radar Squadrons (SAGE).

SAGE was a powerful, albeit expensive system. It was also extraordinarily vulnerable. The combat and direction centers were housed in huge concrete blockhouses, hardened to withstand overpressures of only five pounds per square inch. The advent of Sputnik affected the planning and deployment of the command and control system. Air Force planners realized that Soviet ICBMs could destroy all or part of the SAGE system long before the first of their bombers crossed the Arctic Circle.

Fortunately, the technological achievement of the Soviet SS-6 Sapwood ICBM and Sputnik was matched by an American technological breakthrough of perhaps much greater significance. In the spring of 1958, IBM announced the development of a solid-state computer. Substituting transistors for vacuum tubes, air defense computers could be reduced in size and placed underground in hardened, reinforced concrete facilities. However, ADC plans to construct hardened facilities for SAGE centers were never fulfilled; spending priorities were shifted to develop and deploy American ICBMs.[52]

In March 1961, President John F. Kennedy indicated in his budget message support for a manual back-up system to augment SAGE centers. Speaking before the House Armed Services Committee in April 1961, Secretary of Defense McNamara envisioned adding manual, ground-control intercept capability to augment SAGE centers at radar stations located away from probable target areas.[53]

By the summer of 1961, NORAD was developing plans for what would become known as the Backup Interceptor Control (BUIC) system. Originally, the plan provided for an automated command and control capability for seventy radar stations. Eventually, the list was reduced to thirty in the United States and four in Canada. Some of these sites were planned as master control centers while others were planned as associate centers. Master control centers would assume immediate control of a sector should the regional SAGE direction center be knocked out. Associate centers provided additional redundancy. Site selection criteria focused on vulnerability. A station had to be located at least fifteen miles from an anticipated target. ADC favored locations with good radar coverage along with proximity to interceptor bases. To pay for the program, funds were transferred from the SAGE and FD radar programs.

BUIC implementation was envisioned as a two-phase plan. BUIC I consisted of the manual backup system originally proposed by McNamara. Twenty-seven radar sites were

Searching the Skies: The Legacy of the United States Cold War Defense Radar Program

(1)

(2)

The Evolution to Aerospace Defense (1959–1979)

(1) **A typical SAGE blockhouse. Despite having walls of steel-reinforced concrete six feet thick and interior walls twelve inches thick, the blockhouses were vulnerable to Soviet nuclear-tipped missiles.** (2) **A USAF technician at a SAGE identification radar console selects tracks with a light gun for identification and display on the direction center summary board.** (3) **SAGE equipment at a direction center included this 64 × 64 magnetic core memory.** (4) **Here future SAGE watchstanders are undergoing training at Kessler Air Force Base, Mississippi.** (Official U.S. Air Force photographs courtesy USAF Museum [1] and Air Force History Support Office [2–4].)

(3)

(4)

selected as master or NORAD control centers. Twenty-eight radars acted as associate centers with Ground Control Interception capability. Placing the system in operation simply meant restoring billets that were lost when GCI functions were assumed by the SAGE system. BUIC I reached initial operating capability in December 1962.[54]

At the same time BUIC I was becoming operational, contractors were submitting bids to provide the "brains" for the follow-on automated BUIC II system. BUIC II radar sites would be capable of incorporating data feeds from other radar sectors directly onto their radar screens. In mid-1962, Burroughs Corporation won the contract to provide a military version of its D825 computer to be called the Radar Course Directing Group, AN/GSA-51.[55]

Another back-up command and control program also had its roots during this time. The Airborne Surveillance and Control System (ASACS) was conceived by ADC planners to perform the role of a flying BUIC. Two phases of aircraft were planned with the more sophisticated version to become available in 1970.

Meanwhile, the SAGE system and the primary radar system faced further budget cuts. On December 3, 1962, McNamara recommended closing six SAGE direction command and control centers and seventeen radar stations by mid-1964; President Kennedy approved the measure.[56]

The projected closure of SAGE direction centers sent air defense planners scrambling to prioritize the placement of the thirty-four proposed BUIC II sites. On June 4, 1963, the Air Force approved installation of BUIC II at the first seven sites on the priority list.[57]

During the BUIC II site selection process, at Secretary McNamara's direction, the Air Force undertook a detailed study to examine air defense requirements through 1975. In response, a Continental Air Defense Study (CADS) was completed in May 1963 that called for the replacement of SAGE centers by an improved BUIC and an Airborne Warning and Control System (AWACS) sometime between 1966 and 1975. The improved BUIC sites (BUIC III) foresaw an upgraded AN/GSA-51 capable of integrating surveillance data from ten radars and providing an expanded control capability. The CADS recommended forty-six stations be given this capability. The DoD placed the recommendations on hold to await evaluations on the capabilities of airborne radar operating over land. Eventually, DoD determined that AWACS would not be ready for deployment until after 1970.[58] Work proceeded on the installation of BUIC II and development of BUIC III. In 1966, after the installation of a Burroughs CSA-51 computer system, North Truro Air Force Station (AFS) on Cape Cod became the first ADC installation configured as a BUIC II site. In 1968, North Truro also became the first radar station to be designated a BUIC III installation.[59]

In addition to establishing BUIC sites, extraordinary steps were taken to protect the defense system's control center. Because of the vulnerability of Ent AFB to nuclear attack, planning had begun in 1956 for a more secure command post elsewhere. Former NORAD Commanding General Earle E. Partridge once observed that the two-story

The Evolution to Aerospace Defense (1959–1979)

BIUC II site at North Truro, Massachusetts, August 25, 1965. (Official U.S. Air Force photograph, courtesy National Archives.)

NORAD command building at Ent could be immobilized by a well-aimed bazooka shot, much less by a nuclear blast. To address the problem, architects designed a secure center to be set within man-made caverns in Cheyenne Mountain south of Colorado Springs, Colorado. The Corps of Engineers Omaha District oversaw the massive effort to dig out the caverns. On May 2, 1961, Utah Construction won the bid for the excavation work of the granite mountain. Workers blasted and removed one million tons of granite from inside the mountain. In February 1963, another bid opening placed interior construction work in the hands of Continental Consolidated Corporation. Eleven underground steel buildings were constructed to provide 170,000 square feet of space. To absorb shock waves, each building was mounted on giant steel springs. By February 1966, the "rock" was completed and NORAD began to shift operations from Ent AFB.[60]

In 1966, Secretary of the Air Force Harold Brown proposed a plan that he hoped would overcome McNamara's aversion to air defense. Brown identified survivability, low-altitude detection, responsiveness, and costs as major flaws in the air defense system. In addressing these problems, Brown saw emerging technologies as the key to a more cost-effective and efficient air defense system.

Brown's plan also called for phasing out most military radars around the nation's periphery. Detection duties would be assumed by FAA radars that would feed information into military control centers. To detect low-flying aircraft, Over-The-Horizon-Backscatter (OTH-B) radars were recommended. OTH-B stations would aim powerful radio beams and bounce them off the ionosphere back down on to the earth's surface. In theory, these radars would detect aircraft flying at any altitude at ranges out to 2,000 miles. Brown's plan also pushed for procurement of the AWACS system for survivable command and

Cheyenne Mountain, Colorado, where the NORAD underground Combat Communications Center is located. (Official U.S. Air Force photograph courtesy Air Force History Support Office.)

control capability. Finally, Brown proposed development of the F-12 interceptor to replace aging fighter aircraft in the inventory.[61]

Secretary McNamara approved Brown's plan. He was quick to initiate those portions that cut expenses. Radar stations were closed, and of those that remained, only twenty-two received the BUIC II automated, interception-control capability. By 1968, only radar stations around the nation's perimeter remained in Air Force jurisdiction. All gap-filler radars ceased operations. Interior stations were either closed or turned over to the FAA. However, with the Vietnam War absorbing more defense dollars, McNamara held off on the expenditure portions of Brown's plan. The AWACS and OTH-B program funding was stretched out over several years to support research and development. The F-12 interceptor program eventually was canceled.[62]

The Evolution to Aerospace Defense (1959–1979)

Cheyenne Mountain Status Display Console. (Official U.S. Air Force photograph courtesy National Archives.)

In 1968, ADC became Aerospace Defense Command. However, Aerospace Defense Command fared no better under President Richard M. Nixon's new administration than the ADC had under President Lyndon B. Johnson's. Budget cuts closed down additional radar sites located along the southern perimeter of the country, such as Thomasville, Alabama, and Mount Lemmon, Arizona. By the start of 1970, the number of SAGE centers in the continental United States had been reduced to six: McChord AFB, Washington; Luke AFB, Arizona; Malmstrom AFB, Montana; Duluth International Airport, Minnesota; Hancock Field near Syracuse, New York; and Fort Lee AFS in Virginia. These six remaining SAGE sites still used the vacuum tube Whirlwind II computers.[63]

The ABM Treaty, signed in Moscow in 1972, limited the number of U.S. and Soviet ABM sites to one per country. The treaty signified the ultimate triumph of the offense over defense advocates as national leaders acknowledged that missile defenses were futile. Having adopted the attitude that no defense was possible against missile attacks, national defense strategists determined that continued bomber defenses were also a waste of expenditures. With no new aircraft, interceptor squadrons became antiquated. Many squadrons were disestablished or turned over to the Air National Guard. In 1974, Nike and BOMARC missile defense bases were closed. With the exception of a BUIC III at Tyndall AFB, ADC's BUIC III capability was mothballed. AWACS development continued to be limited to the research and development phase. Once these aircraft finally entered production in the mid-1970s, they were assigned to TAC. OTH-B continued as a research and development project.[64]

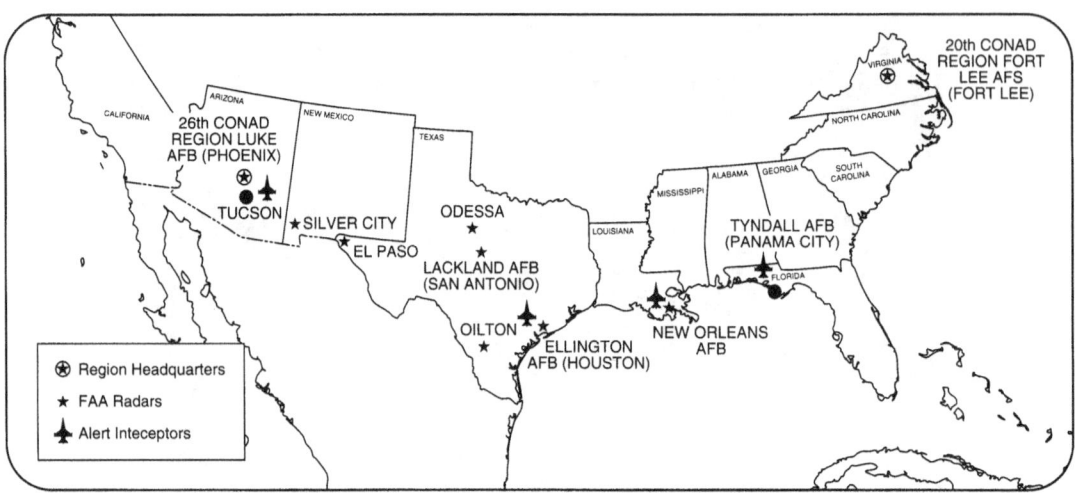

Map of Southern Defense Network. (Map courtesy of Air Combat Command.)

In one region of the country during this period, air defenses received a boost. On October 26, 1971, a Cuban aircraft landed in New Orleans after flying completely undetected through American airspace. Publicity and political pressure from Louisiana Congressman F. Edward Hebert forced the Air Force to redeploy aircraft and radars. Subsequently, the Air Force established the Southeast Air Defense Sector and reopened a radar network along the Gulf coast.[65]

In 1975, reflecting a structural change in organization, ADC's acronym was changed to ADCOM, the Aerospace Defense Command. ADCOM's mission statement called for peacetime protection of air sovereignty and early warning against bomber attack. The command could only provide defense against a limited bomber attack if augmented by units of other commands and services. Because of funding reduction pressures from Congress, in 1977 planners began considering breaking up ADCOM. In 1979, components of ADCOM were turned over to other commands. On October 1, 1979, electronic assets went to the Air Force Communication Service and remaining radar, direction centers, and interceptor forces were transferred to TAC, which became AD-TAC. On December 1, 1979, SAC assumed control of ballistic missile warning and space surveillance facilities. ADCOM was officially disestablished as a major command on March 31, 1980.[66]

Missile Detection and Defense

The Soviet ICBM threat dramatically changed U.S. priorities to building detection and defensive capabilities against ballistic missile attack. Although Sputnik shocked the national psyche, the potential threat of intercontinental ballistic missiles had long been anticipated. Since the German V-2 campaign against England towards the end of World War II, military planners had been working with scientists and engineers to develop an antiballistic missile strategy.

Before the advent of the SS-6 Sapwood and Sputnik, both the Army and the Air Force had been conducting research and development programs leading to an antiballistic missile. The Air Force program, called "Project Wizard," was conceptual in nature. Project Wizard spent millions of dollars in various research labs to develop new technologies to counter the enemy threat. In contrast, the Army program, called "Nike Zeus," was more hardware oriented, building on technology of the earlier Nike Ajax and Nike Hercules antiaircraft missile programs.

In 1958, in the wake of Sputnik, President Eisenhower directed the cancellation of Project Wizard in favor of the Army Nike Zeus program. However, to defend against an attack, the United States needed the capability to detect an attack. Americans feared a nuclear Pearl Harbor, where without warning, nuclear bombs could drop from space, devastating American cities and crippling the military's ability to launch a counterattack. Without the means to defend against such an attack, Americans could only hope that the threat of massive retaliation would deter the Soviet Union from launching such a strike. Early warning would be critical to prepare the nation for the initial blow and allow SAC bombers to get off the ground.

Congress quickly approved funding to construct a Ballistic Missile Early Warning System (BMEWS). Radio Corporation of America (RCA) would develop and build the AN/FPS-49 tracking radars, GE and MIT would design and construct the AN/FPS-50 detection radars, and Western Electric would build the communication systems to connect the radars with command centers. Construction began immediately in the summer of 1958.

BMEWS required building installations at three locations to cover possible flight paths of missiles launched from the Soviet Union. Site I at Thule, Greenland, would host both AN/FPS-49 and AN/FPS-50 radars and receive top construction priority. Providing coverage for most missile approaches from the Eurasian landmass, the Thule site reached initial operating capability in October 1960. Clear, Alaska was selected for Site II to provide warning against missiles launched from the far eastern Siberia region. Initially hosting only AN/FPS-50 detection radars, the Alaskan site began operating in late 1961. Site III, at Fylingdale Moor, Yorkshire, England, was operational in September 1963. At Fylingdale Moor, AN/FPS-49 tracking radars provided coverage of ICBMs launched at the United States from the far western Soviet Union and provided an alert for Europeans if the Soviets launched intermediate range missiles at targets in western Europe.[67]

Construction at the ICBM detection station at Clear began in August 1958. Located eighty miles southwest of Fairbanks, the station consisted of dormitories, administrative buildings, storage warehouses, recreational facilities, radar buildings, transmitter and computer buildings, fuel facilities, and three huge fence antenna components of the AN/FPS-50.

Designed by GE and MIT's Lincoln Laboratory, the three fixed-in-place fence antennas stood 165 feet tall and 400 feet wide. These curved arrays sent two fan-shaped beams at differing angles beyond the earth's atmosphere. When an object passed through the lower-angled beam, the reflected radar pulses were picked up by super-sensitive antennas and passed on to computers that determined the object's position and velocity. When objects passed through the higher-angled second beam, computers received additional information to determine trajectory, speed, impact point, impact time, and launch point. In 1966 a tracking radar was added to the site when Clear received an updated version of the AN/FPS-49. Designated as the AN/FPS-92, this tracking radar featured a movable antenna that locked onto objects identified by the detection radar. This provided additional data to NORAD headquarters.[68]

NORAD received additional contributing sensors. In July 1973, Raytheon won a contract to build a system called "Cobra Dane" on Shemya Island in the Aleutian Islands off the Alaskan coast. Designated as the AN/FPS-108, Cobra Dane replaced AN/FPS-17 and AN/FPS-80 radars placed at Shemya in the 1960s to track Soviet missile tests and to support the Air Force Spacetrack System.[b] Becoming operational in 1977, Cobra Dane also had a primary mission of monitoring Soviet tests of missiles launched from southwest Russia aimed at the Siberian Kamchatka peninsula. This large, single-faced, phased-array radar was the most powerful ever built.[69]

In 1976, the Air Force began operating the Perimeter Acquisition Radar attack Characterization System (PARCS). The story of how the Air Force came to possess this huge, phased-array radar traces its roots back to the 1950s.

In February 1955, the Army contracted Bell Telephone Laboratories to develop an ABM system. This system would be built on the technologies obtained during Nike Ajax and Nike Hercules system development. However, the Nike Zeus system developed by Bell never deployed. Acting on advice that immediate deployment was not technically feasible at an acceptable cost, President Eisenhower decided in May 1959 to maintain Nike Zeus as a research and development program.

By January 1963, the research and development program had evolved into "Nike X." On September 18, 1967, Defense Secretary McNamara acknowledged that ABM defenses could still be overwhelmed by a massive Soviet ICBM attack. However, the emergence of a Chinese nuclear threat could be countered by deploying the Nike X system, renamed the Sentinel, around major metropolitan areas.

---

[b] Spacetrack is an Air Force system that was deployed in the 1960s. It consists of large radars and optical devices positioned globally to monitor objects in Earth orbit. Because satellites play an important role in tracking maritime activities and relaying Naval communications, the Navy established its own space tracking system. The Naval Space Surveillance System consists of a detection fence of nine stations positioned across the southern portion of the United States. The central transmitting station is located at Lake Kickapoo, Texas, with smaller transmitters located at Gila River, Arizona and Lake Jordan in Atlanta, Georgia. Receiving stations were built at San Diego, California, Elephant Butte, New Mexico, Red River, Arkansas, Silver Lake, Missouri, and Hawkinsville and Fort Stewart, Georgia. Data from these stations, along with inputs from Air Force sensors, is sent to Naval Space Surveillance (NAVSPASUR) Headquarters at Dahlgren, Virginia.

The Evolution to Aerospace Defense (1959–1979)

Two views of the U.S. Air Force Ballistic Missile Early Warning System (BMEWS) station located at Clear, Alaska. (Official U.S. Air Force photographs courtesy National Archives.)

Searching the Skies: The Legacy of the United States Cold War Defense Radar Program

On March 14, 1969, the Nixon administration canceled the Sentinel deployment scheme. Instead ABM defense was deployed under the name "Safeguard" to protect America's strategic missile forces. Minuteman missile silos surrounding Grand Forks AFB, North Dakota, and Malmstrom AFB, Montana, would be the first to receive ABM defense.[70]

As a result of the 1972 ABM agreement, the United States completed work only at the site north of Grand Forks. Declared operational in 1975, the Grand Forks ABM site, armed with 100 defending missiles, could provide only a limited defense against the hundreds of warheads that the Soviets could employ. Furthermore, nuclear war scenarios foresaw the radar complexes coming under immediate attack, rendering the intercepting missiles useless. Faced with this futile situation, the Army wanted to operate the system for at least a year and then incorporate the lessons learned for a follow-on system. However, Army plans were cut short on October 2, 1975, when Congress voted to deactivate the site within the following year. Eventually the Air Force assumed operations of Safeguard's Perimeter Acquisition Radar (PAR) and redesignated the site as Cavalier Air Force Station. From its North Dakota location, PARCS provided additional polar coverage to support BMEWS.[71]

BMEWS, along with additional sensors, gave NORAD the capability to warn the National Command Authority of an attack launched from the Soviet Union. However, the

---

PERIMETER ACQUISITION RADAR

The largest of Safeguard's structures, the Perimeter Acquisition Radar (PAR) building, is one of the most solidly constructed buildings in the world. Nearly cubical in shape with dimensions of 204 by 213 feet at the base and rising to over 120 feet, the structure's northern-faced antenna wall sloped away from the ground at a twenty-five degree angle. PAR's phased-array antenna incorporated 6,888 elements, each sending a pulse that would bounce off a target coming over the North Pole. Through a comparison of the reflected signals received back from the incoming object, trajectories were computed. Originally this information was to be passed to the Missile Site Radar. The Missile Site Radar was developed to track the incoming objects and provide guidance information to the interceptor missiles. However, with the shutdown of Safeguard, NORAD determined that the PAR could serve as a fallback sensor to the ballistic missile early warning system and provide data for the Spacetrack system.

Sources: A thorough technical overview completed for the U.S. Army Ballistic Missile Defense Systems Command is found in Bell Laboratories, *ABM Research and Development at Bell Laboratories: Project History, October 1975* (Whippany, NJ: Bell Laboratories, 1975). Construction details were obtained from James H. Kitchens, III, *A History of the Huntsville Division, U.S. Army Corps of Engineers: 1967–1976*, (Huntsville, AL: U.S. Army Engineer Huntsville Division, 1978).

The Evolution to Aerospace Defense (1959–1979)

Soviet Union could attempt to circumvent the warning system using different geographical approaches. The Cuban Missile Crisis of the fall of 1962 was one such attempt. The placement of intermediate range ballistic missiles in Cuba illustrated the vulnerability of the United States to an attack along its unprotected southern border. Only after a high-stakes showdown between the two superpowers, were the missiles removed.

In the wake of the Cuban Missile Crisis, an AN/FPS-85 long-range phased-array radar was constructed at Eglin AFB in Florida. Designed by Bendix Corporation, the radar consisted of a large square transmitter array placed alongside an octangular receiving array mounted on a large structure facing the Gulf of Mexico. The structure hosting the radar burned in 1965, but was rebuilt and placed back in operation in 1969. This radar also served as the main sensor for the Air Force's Spacetrack System and watched the skies over Cuba and the Gulf.[72]

The American triumph of keeping Soviet nuclear launch platforms out of Cuba and at a distance would be short-lived and American defense planners knew it. During the early 1960s, Soviet scientists and engineers worked feverishly to design and build Soviet ballistic missile submarines capable of launching missiles from relatively short distances off America's coastlines. Once again the United States needed the capability to detect incoming missiles to prevent the specter of an atomic sneak attack. In December 1961, the Air Force asked ADC for an evaluation of the capability of FD radars to detect Submarine-Launched Ballistic Missiles (SLBMs). Subsequently, AN/FPS-35 search radars located at Manassas, Virginia, and Benton, Pennsylvania, received modifications

Aerial view of the AN/FPS-85 phased array radar site located at Eglin Air Force Base, Florida.
(Official U.S. Air Force photograph courtesy National Archives.)

and began to be tested during the summer of 1962. During these tests, both radars attempted to track Polaris, Minuteman, Titan, and Thor-Delta missiles launched from Cape Canaveral, Florida. The tests revealed that the AN/FPS-35 had only marginal ability to detect the missile launches.[73] However, using AN/FPS-35 or AN/FPS-24 FD radars to detect SLBMs continued to be considered a viable option given the fiscal constraints imposed on ADC.

Another option to detect SLBMs that was favored by ADC was to procure a series of AN/FPS-49 radars. One of these units had been operating since 1961 at Moorestown, New Jersey, as the original sensor for the Air Force's Spacetrack System. To ADC's disappointment, a study by the Electronic Systems Division at Hanscom AFB, Massachusetts, revealed that using the Moorestown radar for dual use was infeasible.[74]

The long-term vision of ADC planners foresaw SLBM detection as a collateral mission of the OTH-B radar that was still under development. However, ADC could not wait for a system that still was in the research and development stage. In November 1964, desperate to field at least an interim system to warn the nation of a SLBM attack, ADC sought and received permission from the office of the Secretary of Defense to modify existing SAGE system radars.[75]

In the ensuing months, makers of the various SAGE-compatible radar systems submitted proposals on modifications that would enable their products to detect an object of at least two meters in size, at a range of 750 miles, within six seconds after launching. The radar then would continuously track this object within ten seconds of detection and notify the NORAD Combat Operations Center within sixty seconds.

In July 1965, the Air Force selected Avco Corporation for an innovative proposal employing its AN/FPS-26 height-finder radar to detect SLBMs. The modified AN/FPS-26 radar system (redesignated as the AN/FSS-7) was slated for deployment at Point Arena, California; Mount Laguna, California; Mount Hebo, Oregon; Charleston, Maine; Fort Fisher, North Carolina; MacDill AFB, Florida; and Laredo, Texas.[76]

After years of testing and evaluation, the seven-site SLBM detection system became fully operational in 1971. A year later, twenty percent of the surveillance capability of the AN/FPS-85 located at Eglin AFB, Florida, also became dedicated to search for SLBMs.[77]

During the 1970s, the Soviets developed SLBMs that could be launched from greater distances away from the American coastline. For example, the Soviet Navy Delta I class ballistic missile submarine carried the SS-N-8 missile that had a range of over 4,000 nautical miles. This was well beyond the detection capability of either the AN/FSS-7 or the OTH-B radar system being developed.[78] Consequently, the Air Force had to turn to another solution.

The solution was a phased-array warning system to become known as "PAVE PAWS" (Perimeter Acquisition Vehicle Entry Phased-Array Warning System). Originally designed as a two-site system, PAVE PAWS sites were constructed in the late 1970s at

The Evolution to Aerospace Defense (1959–1979)

An exterior view of the PAVE PAWS located at Beale Air Force Base in California. (Official U.S. Air Force photograph courtesy National Archives.)

Otis AFB, Massachusetts, and Beale AFB, California. From a distance, the PAVE PAWS structure looked like a three-sided pyramid with a flattened top. On the two seaward faces of the pyramid, Raytheon installed the AN/FPS-115 with its phased-array antenna. Thirty meters in diameter and consisting of 2,000 elements, each antenna could detect objects launched as far away as 3,000 miles. The Otis site became operational in 1979 and the Beale site became operational a year later.

A contract for two more continental PAVE PAWS sites was awarded in 1984. An AN/FPS-115 at Robins AFB, Georgia, became operational in 1986 and another unit at Eldorado AFS, Texas, was activated in 1987. Additional AN/FPS-115 PAVE PAWS radars were installed in the 1990s at BMEWS sites at Thule, Greenland, and Fylingdale Moor,

Searching the Skies: The Legacy of the United States Cold War Defense Radar Program

Interior view of the PAVE PAWS located at Beale Air Force Base in California. (Official U.S. Air Force photograph courtesy National Archives.)

England, to assume the ICBM detection mission. As PAVE PAWS sites in the United States were activated, the older AN/FSS-7 radars were phased out, except for the MacDill AFB site that continued to provide additional coverage over Cuba.[79]

Spacetracking and missile detection functions of the former Aerospace Defense Command were assumed by SAC in 1980. Control of these facilities became an Air Force Space Command responsibility with the activation of that command on September 1, 1982.

CHAPTER 3

AIR DEFENSE REVITALIZED (1979–1994)

Looking to the Future

In March 1983, President Ronald W. Reagan announced his controversial Strategic Defense Initiative (SDI). Although SDI focused on ballistic missile defense, the implications for air defense were profound. Air Force Director of Plans Major General John A. Shaud observed that a ballistic missile defense system would have to be complemented by an air defense system. He stated "If you're going to fix the roof, you don't want to leave the doors and windows open."[80] Thus the efforts rooted in the 1966 Brown Plan would finally come to fruition during the 1980s and early 1990s.

Rebuilding the Network

Steps were taken to improve the air defense warning system long before President Reagan announced his Strategic Defense Initiative. The absorption of ADCOM into TAC in 1980 came during the transition to a system that had been envisioned by the Brown Plan over a decade earlier. The DoD and the FAA had been negotiating throughout the 1970s for the FAA to assume control of most tracking duties as part of a proposed Joint Surveillance System. To create the JSS, during 1979 and 1980 TAC closed down twenty-seven SAGE radar sites. Some of these sites were retained to become FAA-operated JSS sites. In other cases, the former ADCOM sites were placed in caretaker status. At some operational FAA sites, a small Air Force detachment arrived to install and operate a height-finder radar. Radars built for the FAA did not have a height-finding capability.

In the early 1980s, when the JSS project was completed, the JSS operated forty-six long-range radar sites. Thirty-one of the sites had FAA-operated search radars and Air Force-manned height-finder radars. Five sites had FAA radars that simply provided a data tie to one of the SAGE Regional Control Centers (RCC). The ten remaining long-range radar sites were operated by the military. Six of those sites were operated by the Air Force. The Oceana Naval Air Station site in Virginia was jointly operated by the Navy and Air Force. Contractors operated a radar at Lake Charles, Louisiana, and Civil Service personnel operated a radar at Point Arena, California. The remaining DoD site, at Cudjoe Key, Florida, used a radar that was flown within an aerostat balloon.[81]

Initially, these forty-six radar sites provided data feeds to the six remaining SAGE RCCs. During 1983, these six RCCs were replaced by four Region Operation Control Centers (ROCCs) that operated as part of the JSS.

Map of the JSS System. (Map courtesy of Air Combat Command.)

Within the continental United States, ROCCs were located at Griffiss AFB, New York; March AFB, California; Tyndall AFB, Florida; and McChord AFB, Washington. Additional ROCCs were located in Alaska and Hawaii. Canada established two ROCCs collocated at North Bay, Ontario. Like the RCCs of the previous SAGE system, these ROCCs were "soft," meaning that they were vulnerable to nuclear attack. Consequently, the centers were designed for peacetime use to detect, track, and identify intruding aircraft, and if needed, to deploy and direct interceptor aircraft to challenge an intruder. If the United States became threatened by a direct attack, the ROCCs had the capability to transfer command and control function to an airborne AWACS aircraft.[82]

The ROCCs saved millions of dollars over the previous system. In contrast to the Whirlwind II computer that occupied a half-acre of space within the old SAGE blockhouses, the more capable computers for the ROCCs occupied the space of a vending machine. ROCC computer components were also more readily available than the SAGE vacuum tubes that often had to be procured from eastern bloc countries. Not only was the new computer more capable, but it allowed ROCC watchstanders to perform their duties in normal room temperatures. To maintain the earlier Whirlwind II computer, huge air conditioning units kept the vacuum tubes and the watchstanders in a cool environment.[83]

Air Defense Revitalized (1979–1994)

Tyndall NORAD Regional Control Center (RCC) exterior and interior. (Official U.S. Air Force photographs courtesy of Department of Defense Still Media Records Center.)

Airborne Warning and Control System (AWACS) aircraft became a key component of United States air defense in the 1980s. (Official U.S. Air Force photograph courtesy National Archives.)

As these ROCCs became operational, TAC made additional cuts; deactivated remaining continental radar squadrons and disestablished detachments that had operated height-finder radars at the FAA-operated joint-use sites. By 1987, the four ROCCs relied mostly on data-feeds from the FAA JSS radars. Technological advances allowed for these equipment and manpower reductions and improved U.S. detection and tracking capabilities. For example, in July 1988, Westinghouse Electric Corporation received a contract to build forty ARSR-4 radar systems for installation at JSS sites around the periphery of the nation. A 3-D radar, the ARSR-4 was the first radar truly capable of meeting both the air traffic control and air defense requirements for the FAA and Air Force. With the installation of these units during the mid-1990s, radars of the 1950s and 1960s vintage finally could be retired.[84]

In addition to establishing the JSS system, the Air Force implemented another of the 1966 Brown recommendations: construction of east and west coast OTH-B sites. In 1975, GE Aerospace received the contract to build a prototype OTH-B system. The transmitter site was built at Moscow AFS, Maine, and the receiver site was constructed at Columbia AFS, Maine. Initial testing occurred in 1980. With successful test trials, GE Aerospace received a contract in 1982 to build a full-scale model of the system designated as the AN/FPS-118. The east coast system achieved limited operational capability in 1988. Two years of testing and evaluation on the east coast yielded improvements for the west coast

Air Defense Revitalized (1979–1994)

system that was undergoing construction. The Air Force accepted the east coast system from the contractor in April 1990.

The west coast OTH-B operations center was located at Mountain Home AFB in Idaho. The transmitter was placed at Christmas Valley, Oregon, and the receiver was erected at Tule Lake near Alturas, California. The west coast system was accepted by the Air Force at the end of 1990.

In March 1991, a diminished threat led to a recommendation to scrap the whole system. However, the Air Force decided to conduct limited operations on the east coast and preserve the system in a maintenance status on the west coast.[85]

Chapter 4

Epilogue

Much remains of the air and aerospace detection, command, and control systems built during the Cold War. Although only a fraction of the radar stations built during the 1950s and 1960s remain in military hands, many are still operational under FAA control. However, the FAA is in the process of completing its modernization program to replace Air Force 1960s vintage FPS model radars. At former ADC sites, the radars have been removed and the facilities have been converted to perform new functions. Many sites, especially in remote locations, simply have been abandoned.

The blockhouses that once hosted SAGE centers remain intact at many locations, although the Whirlwind II computers and command consoles have long been removed. The four ROCCs built during the 1980s remain intact and operational. The intruding aircraft in the 1990s represent a different threat; attempting to smuggle illegal drugs into the country.

The BMEWS system will remain intact for the foreseeable future as long as more countries gain the capability to launch ballistic missiles. Cheyenne Mountain, Colorado, still serves as the nerve center for North America's missile tracking sensors.

Historians will long argue what brought about the demise of the Soviet Union and why World War III never was fought. While one school argues that the Soviet system collapsed under its own weight of inefficiency, another school vigorously contends that American military vigilance significantly contributed to the Soviet demise.

Nuclear deterrence, it is argued, eliminated direct military confrontation as an option for the Soviets. If such is the case, then the role of the thousands of men and women who operated the radar stations and command centers during the Cold War cannot be overlooked. They contributed to the deterrence in two ways. First, by being able to direct interceptor forces against intruding aircraft, the air defenders reduced the opponent's confidence level for mission success. Second, and more importantly, the warning provided by the air defense and later missile defense warning sensors gave America's nuclear forces the forewarning necessary to deliver a devastating retaliatory blow.

When viewing the hundreds of abandoned air defense structures dotting the American landscape, one should reflect on the roles of the thousands of men and women who operated the air defense systems. Part of their legacy is their contribution to the United States' triumph in the Cold War.

ENDNOTES

1. Kenneth Schaffel, "The U.S. Air Force's Philosophy of Strategic Defense: A Historical Overview," in *Strategic Air Defense*, Stephen J. Cimbala ed., (Wilmington, DE: Scholarly Resources Inc.. 1989), p. 5; *A Handbook of Aerospace Defense Organization, 1946–1986,* (Peterson Air Force Base, CO: Office of History Air Force Space Command, 1987), p. 1.
2. Sean S. Swords, *Technical History of the Beginnings of Radar* (London: Peter Peregrinus, 1986), pp. 112–16. A detailed account of the Army effort is provided in Harry M. Davis, *History of the Signal Corps Development of the U.S. Army Radar Equipment: Research and Development, 1918–1937 (Part I)* (Washington: Office of the Chief Signal Corps Officer, Historical Section Special Activities Branch, 1944); and Dulany Terrett, *United States Army in World War II: The Signal Corps—The Emergency (To December 1941)* (Washington: Office of the Chief of Military History, 1956).
3. Schaffel, *Strategic Air Defense,* pp. 7–9; Norman Friedman, *Naval Radar* (Annapolis, MD: Naval Institute Press, 1981), pp. 80–82, 88n. Two works covering Naval Research Laboratory radar work include David K. Allison, *A New Eye for the Navy: The Origin of Radar at the Naval Research Laboratory, NRL Report 8466* (Washington: U.S. Government Printing Office, 1981), and Louis A. Gebhard, *Evolution of Naval-Electronics and Contributions of the Naval Research Laboratory, NRL Report 8300* (Washington: U.S. Government Printing Office, 1979). Bell Laboratories contributions are covered in M.D. Fagen, ed., *A History of Engineering and Science in the Bell System: National Service in War and Peace (1925–1975)* (New York: Bell Telephone Laboratories, 1978).
4. Edwin T. Layton, *"And I Was There": Pearl Harbor and Midway—Breaking the Secrets* (New York, NY: William Morrow and Company, Inc. 1985), p. 308.
5. Friedman, *Naval Radar,* pp. 87–88.
6. Schaffel, *Strategic Air Defense,* pp. 10–12. See also Chapter 1 in Schaffel's *The Emerging Shield: The Air Force and the Evolution of Continental Air Defense, 1945–1960* (Washington, DC: Office of Air Force History, 1991) and *A Handbook of Aerospace Defense Organization,* pp. 4–5.
7. Merrill I. Skolnik, "Fifty Years of Radar," *Proceedings of the IEEE* (February 1985), pp. 183–84. For a detailed look at Radiation Laboratory activity see *Five Years at the Radiation Laboratory* (Boston: 1991 IEEE MIT-S International Microwave Symposium 1991).
8. *A Handbook of Aerospace Defense Organization,* p. 4.
9. Schaffel, *The Emerging Shield,* pp. 47–48; Richard F. McMullen, "The Aerospace Defense Command and Anti-Bomber Defense, 1946–1972," ADC Historical Study No. 39, 1973, p. 2.
10. Richard F. McMullen, "Radar Programs for Air Defense, 1946–1966," ADC Historical Study No. 34 (1966), pp. 1–2.
11. Schaffel, *The Emerging Shield,* pp. 54–55; McMullen, "Radar Programs," p. 3; McMullen, "Antibomber Defense," pp. 2–5.
12. Schaffel, *The Emerging Shield,* pp. 61–63; McMullen, "Radar Programs," pp. 5–8; McMullen, "Antibomber Defenses," pp. 14–15; "Hearings before the Subcommittee of the House Committee on Appropriations for the Military Establishment," February 17, 1947, p. 629.
13. Schaffel, *The Emerging Shield,* pp. 67–68; McMullen, "Radar Programs," pp. 9–10; McMullen, "Antibomber Defense," pp. 19–20.
14. McMullen, "Radar Programs," p. 8; Schaffel, *The Emerging Shield,* p. 77; McMullen, "Antibomber Defense," p. 22.
15. Schaffel, *The Emerging Shield,* p. 78; McMullen, "Radar Programs," pp. 12–13, 16–18; McMullen, "Antibomber Defenses," pp. 30–31.
16. Schaffel, *The Emerging Shield,* pp. 91–93.
17. Schaffel, *The Emerging Shield,* pp. 95–96; McMullen, "Radar Programs," pp. 18–19.
18. Schaffel, *The Emerging Shield,* pp. 107–110; McMullen, "Radar Programs," pp. 23–24.
19. McMullen, "Radar Programs," p. 26.
20. Schaffel, *The Emerging Shield,* pp. 120–21.
21. Schaffel, *The Emerging Shield,* pp. 129, 134–35; Lieutenant Colonel Steve Moeller, *Vigilant and Invincible: The Army's Role in Continental Air Defense, 1950–1974* (Columbus, OH: Ohio State University Master's Thesis, 1992), pp. 21–27.
22. McMullen, "Radar Programs," pp. 27–28.

23. Ibid., pp. 30–32, 34.
24. Schaffel, *The Emerging Shield,* pp. 140–41.
25. McMullen, "Radar Programs," pp. 44–45.
26. McMullen, "Radar Programs," p. 47; Schaffel, The Emerging Shield, pp. 155–56.
27. Schaffel, *The Emerging Shield,* pp. 144–45, 150; "The Truth About Our Air Defense," Air Force (May 1953), p. 29; McMullen, "Antibomber Defense," pp. 50–53.
28. Schaffel, *The Emerging Shield,* p. 205; Eva C. Freeman, ed., *MIT Lincoln Laboratory: Technology in the National Interest* (Lexington, MA: Lincoln Laboratory, Massachusetts Institute of Technology, 1995), pp. 15, 17.
29. Schaffel, *The Emerging Shield,* pp. 150–51, 169–71.
30. Richard Morenus, *The DEW Line: Distant Early Warning, The Miracle of America's First Line of Defense* (New York, NY: Rand McNally and Company, 1957), Chapter 2; Schaffel, *The Emerging Shield,* pp. 174–77.
31. Schaffel, *The Emerging Shield,* pp. 185–86.
32. Ibid., pp. 190–191, 193; McMullen, "Antibomber Defense," pp. 55, 59–64.
33. McMullen, "Radar Programs," pp. 57–58; Schaffel, *The Emerging Shield,* p. 193.
34. Schaffel, *The Emerging Shield,* p. 194; Frederic H. Smith, Jr. "How Air Defense Is Part of the Great Deterrence," *Air Force* (June 1956), pp. 90–91, 93.
35. McMullen, "Radar Programs," pp. 57–58.
36. Virge Jenkins Temme, et al., *Historical and Architectural Documentation Reports of North Truro Air Force Station, North Truro, Massachusetts* (Langley AFB, VA: Headquarters, Air Combat Command, 1995), pp. 7–9; Morenus, *DEW Line,* Chapter 1.
37. Schaffel, *The Emerging Shield,* pp. 203–5; Freeman, *MIT Lincoln Laboratory,* pp. 25, 27.
38. McMullen, "Radar Programs," pp. 77–78; Schaffel, *The Emerging Shield,* pp. 214–16; DEW Line construction is discussed in Charles Corddry, "How we're building the world's biggest Burglar Alarm," *Air Force* (June 1956); and Howard La Fay, "DEW Line, Sentry of the Far North," *The National Geographic Magazine,* (July 1958).
39. McMullen, "Radar Systems," pp. 70–71.
40. Ibid., pp. 59–61, 71.
41. Ibid., pp. 87, 105, 126.
42. *Historical Data of the Aerospace Defense Command, 1946–1973,* p. 78; Schaffel, *The Emerging Shield,* p. 222.
43. Letter dated October 31, 1995, from NORAD historian Dr. Thomas Fuller to author.
44. Richard F. McMullen, "Air Defense and National Policy: 1958–1964," (ADC Historical Study No. 26, 1965), p. 45.
45. McMullen, "Radar Programs," p. 231.
46. Russell F. Weigley, in *The American Way of War: A History of United States Military Strategy and Policy* (New York, NY: 1973), argues that offensive doctrine has prevailed throughout American military history.
47. Schaffel, *The Emerging Shield,* p. 261.
48. McMullen, "Radar Programs," pp. 138–39.
49. Ibid., pp. 140–42, 152, 167–68.
50. Ibid., pp. 155–56, 180–81.
51. Robert Frank Futrell, *Ideas, Concepts, Doctrine: Basic Thinking in the United States Air Force* (Maxwell Air Force Base, Montgomery, AL: Air University Press, [1971] 1989), pp. 532–33.
52. Richard F. McMullen, "Command and Control Planning: 1958–1965," ADC Historical Study No. 35, (1965), pp. 1–11.
53. Ibid., pp. 15–16.
54. Ibid., pp. 26–27.
55. Ibid., p. 38.
56. Ibid., pp. 43–45.
57. Ibid., pp. 50–55.
58. Ibid., pp. 57–64.
59. "Declaration of Excess Real Property North Truro Air Force Station, North Truro Massachusetts, 18 June 1984," p. 5.
60. *The Federal Engineer Damsites to Missile Sites: A History of the Omaha District U.S. Army Corps of Engineers* (Omaha, NE: U.S. Army Engineer District Omaha, 1984), pp. 199–213; Stanley L. Englebardt, *Strategic Defenses* (New York, NY: Thomas Y. Crowell Company, 1966), pp. 136–39; "Cheyenne Mountain Complex Chronology" (July 1994) provided by United States Space Command.

Endnotes

61. Owen E. Jensen, "The Years of Decline: Air Defense from 1960 to 1980," in Strategic Air Defense, Stephen J. Cimbala, ed. (Wilmington DE: Scholarly Resources Inc. 1989), pp. 32–35.
62. Ibid., p. 38.
63. Headquarters Aerospace Defense Command Press Release (undated) located at the Air Force Museum Research Center collection File L2 Command Air Defense.
64. Jensen, "Years of Decline," pp. 23–24, 38–39.
65. Ibid., pp. 39–40; Claude Witze, "The Gaps in Our Defense," *Air Force* (March 1972), pp. 33–39.
66. Jensen, The Years of Decline," pp. 40–41; *A Handbook of Air Defense Organization,* p. 6.
67. Englebardt, *Strategic Defenses,* pp. 107–8; *A Handbook of Aerospace Defense Organization: 1946–86,* pp. 22, 24.
68. Denfield, *Cold War in Alaska,* pp. 40–41; Englebardt, *Strategic Defenses,* pp. 108–110; *Jane's,* p. 79; Schaffel, *The Emerging Shield,* pp. 257–60.
69. *Jane's Radar and Electronic Systems,* 6th edition, Bernard Blake, ed. (1994), p. 78.
70. Donald Baucom, *The Origins of SDI, 1944–1983* (Lawrence, KS: University Press of Kansas, 1992) provides an excellent integration of the political and technical aspects of the program.
71. James H. Kitchens, III, *A History of the Huntsville Division, U.S. Army Corps of Engineers, 1967–1976* (Huntsville, AL: U.S. Army Corps of Engineers Huntsville Division, 1978), pp. 111–12; Letter dated May 2, 1996 from Mandy Whorton, Argonne National Laboratory to Virge Jenkins-Temme, USACERL.
72. *Jane's,* p. 80.
73. McMullen, "Radar Programs," pp. 171–74.
74. Ibid., pp. 188–91. The Moorestown AN/FPS-49 radar was phased out in 1969 once the AN/FPS-85 phased array radar at Eglin, Air Force Base assumed Spacetrack duties. See "Headquarters Aerospace Defense Command News Release 10-28-209-APM," at Research Center, Air Force Museum, Wright-Paterson Air Force Base, folder L2 Command Air Defense.
75. McMullen, "Radar Programs," pp. 205–6.
76. Ibid., pp. 224–25.
77. *Jane's,* p.31.
78. *Understanding Soviet Naval Developments,* 3rd ed. (Washington, DC: Office of the Chief of Naval Operations, 1978), p. 73.
79. *Jane's,* p. 81; "PAVE PAWS Radar System," *United States Air Force Fact Sheet* (Peterson, AFB, CO: Office of Public Affairs, Air Force Space Command, 1992).
80. Rick Atkinson, "Air Defense for Continental U.S. Is Coming Back Into Vogue," *The Washington Post* (August 25, 1984), p. A4.
81. From "JSS documents," provided by Air Combat Command History Office.
82. *Jane's,* p. 32.
83. Marilyn Silcox, "Southeast ROCC Marks Beginning of New Air Defense Era," National Defense (July–August 1984), pp. 42–43, 46; Rick Atkinson, "Air Defense for Continental U.S. Coming Back Into Vogue," *The Washington Post* (August 25, 1984), p. A4.
84. *Jane's,* p. 85.
85. Ibid., pp. 82–83.

ARCTIC DUTY 1954

The following piece, written by a member of the Alaskan Air Command Public Affairs Office, captures the flavor of radar duty in the far north during the early Cold War.

MERRY CHRISTMAS AMERICA
by A/1C Casey Wondergem

It's Christmas Eve. The Wind outside our radome is singing in the 60 degree below night . . . like a mighty chorus caroling in the darkness. The moonlight sparkles on the crisp carpet of snow . . . but it's a carpet which leads to "nowhere." For our world is a nest of huts, perched atop a lofty, snow-covered mountain peak . . . a world of ice-crusted cat-walks and tramways. It's a quiet world and the stillness is seldom broken except by the lonely howl of our pet husky or the welcoming roar of a visiting bush plane.

No, we're not begging for compassion. For we're a busy lot, and hence a happy lot. Our mission is one of eternal vigilance . . . our scopes must be scanned around the clock. We're sitting on your front-doorstep, America, ever ready to sound the alarm on our first intruder. Although we may dream of more luxurious places and "tropical pleasures," we're content with our lot. We're proud of our job . . . proud of each other. We've a feeling for the responsibility you've entrusted to us and we've got the stamina to stick it out.

Yes, it's Christmas Eve here in the Bering Sea and the winter night falls fast on these far reaches of the Arctic. There is no brightly lit Christmas trees outside to add warmth to the countryside . . . in fact, you cannot find a tree within a hundred miles. We're alone, but not lonely . . . we're isolated but not discontented. We understand the

(Official U.S. Air Force photograph courtesy National Archives.)

Arctic Duty, 1954

purpose of our vigil and appreciate the import of our responsibility.

We're not particularly overjoyed or enthralled with the "luxury" of this mountain peak, but we thought perhaps you might be interested in learning of the secret to our contentment. We're happy, despite it all, because we've learned how to live together and work together toward a common goal. We've learned to understand one another and resolve our difficulties . . . to share each other's joys as well as problems. Regardless of race, creed or color, we're all "brothers" as well as scope-scanners at this radar site.

Yes, Merry Christmas, America! They've dropped us a tree to augment our Yuletide cheer . . . and we've got the tinsel, and lights, and gayly wrapped gifts to make our spirit burn brightly through the Christmas season. So think of us, 'cause we'll be thinking of you. And may our common prayer be that the harmony and brotherhood which exists at this lonely outpost may someday flourish throughout the world—"On earth peace, good will towards men."

Source: National Archives and Records Administration Still Picture Branch, Air Force binder 342B 05002.

Part II

Systems Overview

Radar Systems Classification Methods

During World War II, each service used its own method to designate its electronic radar/tracking systems. For example, Army radars were classified under the initials SCR, which stood for "Signal Corps Radio." Different designations for similar systems confused manufacturers and complicated electronics procurement. In February 1943, a universal classification system was implemented for all services to follow, ending the confusion. To indicate that an electronic system designation followed the new universal classification, the letters "AN," for Army-Navy, were placed ahead of a three-letter code. The first letter of the three-letter code denoted the type of platform hosting the electronic device, for example: A=Aircraft; C=Air transportable (letter no longer used starting in the 1950s); F=Fixed permanent land-based; G=General ground use; M=Ground mobile; S=Ship-mounted; T=Ground transportable. The second letter indicated the type of device, for example: P=Radar (pulsed); Q=Sonar; R=Radio. The third letter indicated the function of the radar system device, for example: G=Fire control; R=Receiving (passive detection); S=Search; T=Transmitting. Thus an AN/FPS-20 represented the twentieth design of an Army-Navy "Fixed, Radar, Search" electronic device.

World War II Radars

This section describes the World War II vintage radars that saw service during the Cold War. The systems are listed in numerical order, bypassing the three-letter code. During World War II, search and height-finder radars became components of America's electronic arsenal. The function of the search radar was to detect and obtain a line of bearing on an aircraft. Early models such as the SCR-270 and 271 looked like large bedsprings. Later designs, such as the AN/CPS-5 looked like a large oval dish. Search radars generally rotated full circle around a central axis. In contrast to the rotating search radar antenna, the horizontally mounted height-finder radar focused on the tracked aircraft's reported bearing. The radar antenna dish then scanned up and down to provide the operators with the estimated height of the aircraft.

AN/TPS-1B, 1C, 1D

Bell Telephone Laboratories developed this radar that subsequently was produced by the Western Electric Company. A crew of two could operate the radar. The 1B model could detect bombers at 10,000 feet at a distance of 120 nautical miles. The height detection and range on the 1C and 1D models exceeded those of the 1B. The transmitter sent its pulse at an L-band frequency between 1220 to 1280 megahertz (MHz). This long-range search radar was used in the temporary Lashup system beginning in 1948.

AN/CPS-4

Developed by MIT's Radiation Laboratory, this height-finding radar was nicknamed "Beaver Tail." The radar was designed to be used in conjunction with the SCR-270 and SCR-271 search sets. The CPS-4 required six operators. This S-band radar, operating in the 2700 to 2900 MHz range, could detect targets at a distance of ninety miles. The vertical antenna was twenty feet high and five feet wide. This radar was often paired with the AN/FPS-3 search radar during the early 1950s at permanent network radar sites.

AN/CPS-5

Bell Telephone Laboratories and General Electric developed this search radar. General Electric began producing sets in January 1945. Designated as a transportable medium-range search radar, the unit was ideal for use in the Lashup system in conjunction with the AN/TPS-10 height-finder radar. It could be operated with a crew of ten. Some of these units remained to serve in the first permanent network. Designed to provide a solid search of up to 60 miles at 40,000 feet, the radar often had success tracking aircraft as far as 210 miles away.

AN/CPS-5

AN/CPS-6, 6A, 6B

The AN/CPS-6 was developed during the later stages of World War II by the Radiation Laboratory at MIT. The first units were produced in mid-1945. General Electric developed and produced the A-model and subsequent B-model at a plant in Syracuse, New York. The unit consisted of two antennas. One of the antennas slanted at a forty-five degree angle to provide the height-finder capability. Initially, the radar was designed to detect fighter aircraft at 100 miles and 16,000 feet. The radar used five transmitters that operated at S-band frequencies ranging from 2700 to 3019 MHz. It took twenty-five people to operate the radar. An AN/CPS-6 radar was installed as part of the Lashup system at Twin Lights, New Jersey, in 1949 and proved capable of detecting targets at ranges of eighty-four miles. The first units of the follow-on 6B radar set were ready for

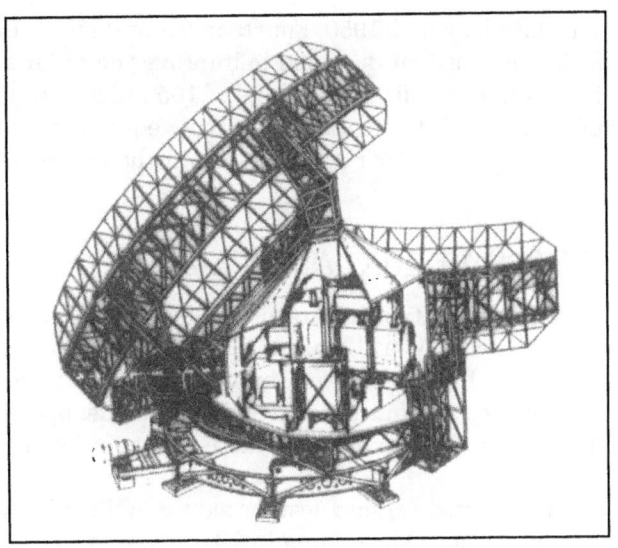

AN/CPS-6

CHANGING TERMINOLOGIES

Terms describing radar characteristics evolved during the Cold War. For example, frequency was once measured in megacycles per second (mc/s). Now the term megahertz (MHz) conveys the same measurement.

There are different letter-code systems in existence to identify frequency bands. This study uses the U.S. military system that originated during World War II. This standard has been adopted by the Institute of Electrical and Electronic Engineers. However, the North Atlantic Treaty Organization (NATO) has its own letter code system that is used in numerous reference books. The following chart is provided for easy cross-reference:

U.S.	NATO	Frequency (MHz)	Wavelength (cm)
VHF	A	30–300	300–100
UHF	B/C	300–1000	100–30
L	D	1000–2000	30–15
S	E/F	2000–4000	15–7.5
C	G/H	4000–8000	7.5–3.75

installation by mid-1950. Fourteen 6B units were used within the first permanent network. A component designed to improve the radar's range was added in 1954. Initial tests showed the 6B unit had a range of 165 miles with an altitude limit of 45,000 feet. One radar unit and its ancillary electronic equipment had to be transported in eighty-five freight cars. The Air Force phased out the 6B model between mid-1957 and mid-1959.

AN/TPS-10, 10A / AN/FPS-4

MIT's Radiation Laboratory developed and produced the first version of this radar near the end of World War II. Zenith produced the A-model sets in the post-war period. The vertically mounted antenna was three feet wide and ten feet long. Two operators were needed to run the set. The initial model operated at a frequency of 9000 to 9160 MHz and had a maximum reliable range for bombers of 60 miles at 10,000 feet.

An updated version designated the AN/FPS-4 was produced by the Radio Corporation of America (RCA) beginning in 1948. Some 450 copies of this and the trailer-mounted AN/MPS-8 version were built between 1948 and 1955.

Early Cold War Search Radars

Early Cold War search radars essentially were advanced or improved versions of World War II era sets. In some cases, the performance of the new sets fell short of expectations.

AN/FPS-3, 3A

The AN/FPS-3 was a modified version of the AN/CPS-5 long-range search radar. The first units came off the Bendix production line and were ready for installation in late 1950. Forty-eight of these L-band units were used within the first permanent network. The AN/FPS-3B incorporated an AN/GPA-27, which increased the search altitude to 65,000 feet. Installation of these modifications began in 1957.

AN/FPS-5

The AN/FPS-5 was a long-range search radar produced in the early 1950s by Hazeltine. Deployment was limited.

AN/FPS-8

The AN/FPS-8 was a medium-range search radar operating on the L-band at a frequency of 1280 to 1380 MHz. Developed in the 1950s by General Electric, over 200 units of this radar were produced between 1954 and 1958. Variants of this radar included the AN/GPS-3 and the AN/MPS-11.

AN/FPS-3

AN/FPS-8

AN/FPS-10

This unit was essentially a stripped down version of the AN/CPS-6B. Thirteen of these units served within the first permanent network.

SAGE System Compatible Search Radars

Various manufacturers began design work on compatible search radars for SAGE systems in the mid-1950s in conjunction with the development of the SAGE Command and Control System. Because Project LAMPLIGHT indicated radar vulnerability to electronic countermeasures, the Air Force developed a series of radars that could shift frequency. These frequency-diversity (FD) radars included the AN/FPS-24, AN/FPS-27, and AN/FPS-35.

AN/FPS-7, 7A, 7B, 7C, 7D

In the mid-1950s, General Electric developed a radar with a search altitude of 100,000 feet and a range of 270 miles. This radar was significant in that it was the first stacked-beam radar to enter into production in the United States. Designed to operate in the L-band at 1250 to 1350 MHz, the radar deployed in late 1959 and the early 1960s. The AN/FPS-7 was used for both air defense and air traffic control in New York, Kansas City, Houston, Spokane, San Antonio, and elsewhere. In the early 1960s, a modification called AN/ECP-91 was installed to improve its electronic countermeasure (ECM) capability. About thirty units were produced.

AN/FPS-20, 20A, 20B

This Bendix-built radar was an AN/FPS-3 search radar with an AN/GPA-27 installed. Designed to operate in the L-band frequencies of 1250 to 1350 MHz, the radar had a range of over 200 miles. By the late 1950s this radar dominated the United States radar defense net. Deployment continued into the early 1960s. In June 1959, Bendix received a contract to provide private industry's MK-447 (the same as the military's AN/GPA-103)

AN/FPS-20

and MK-448 (AN/GPA-102) anti-jam packages to the radars. With the addition of these packages, the Air Force redesignated the radars. The AN/FPS-20A with the AN/GPA-102 became the AN/FPS-66 and the AN/FPS-20A with the AN/GPA-103 became the AN/FPS-67. Over 200 units were built.

AN/FPS-24

AN/FPS-24

General Electric built an FD search radar designed to operate in the Very High Frequency (VHF) at 214 to 236 MHz. There were problems with this radar at the test site at Eufaula, Alabama, in 1960. These problems required many modifications. Additional problems occurred when deployment was attempted in 1961. When the radar finally deployed, bearing problems often occurred due to the eighty-five ton antenna weight. Twelve systems were built between 1958 and 1962.

AN/FPS-27, 27A

Westinghouse built an FD search radar designed to operate in the S-band at 2322 to 2670 MHz. The radar was designed to have a maximum range of 220 nautical miles and search to an altitude of 150,000 feet. System problems required several modifications at

AN/FPS-27

the test platform located at Crystal Springs, Mississippi. Once these problems were solved, the first of twenty units in the continental United States became operational at Charleston, Maine, in 1963. The last unit was installed at Bellefontaine, Ohio, a year later. In the early 1970s, AN/FPS-27 radar stations that had not been shutdown received a modification (solid state circuitry replacing vacuum tubes) that improved reliability and saved on maintenance costs.

AN/FPS-28

Raytheon designed this search radar to operate at 410 to 690 MHz. A test unit was placed at Huoma Naval Air Station (NAS) in Louisiana.

AN/FPS-30

Bendix built this long-range search radar that operated in the L-band.

AN/FPS-31

Designed by Lincoln Laboratory, this huge radar was designed to be compatible with the SAGE system. A prototype was built at Jug Handle Hill in West Bath, Maine. The antenna was 120 feet wide and 16 feet high. Operations began in October 1955. After a period of unexpected clutter, it was determined that the radar received echoes from the aurora borealis (Northern Lights) and this hindered tracking. Although this model was never mass-produced for active use, lessons learned from this radar would continue supporting SAGE system research and development.

AN/FPS-35

This Sperry-built FD long-range search radar was designed to operate at 420 to 450 MHz. It was first deployed in December 1960, but problems hampered the program. Four of these units were operational in 1962. The system suffered frequent bearing problems as the antenna weighed seventy tons.

AN/FPS-64, 65, 66, 67, 67A, 72

These radars were modified versions of the Bendix AN/FPS-20 search radar. See the AN/FPS-20 entry.

AN/FPS-87A

Bendix built this long-range L-band search radar that was based on the AN/FPS-20. See the AN/FPS-20 entry.

AN/FPS-88

General Electric produced this updated version of the AN/FPS-8 radar in the late 1960s. The AN/FPS-88 operated in the L-band at 1280 to 1380 MHz and featured some ECM capability.

AN/FPS-91

This radar was another version of the AN/FPS-20 search radar produced by Bendix. See the AN/FPS-20 entry.

AN/FPS-93

Raytheon modified the AN/FPS-20 radar to create this radar. See the AN/FPS-20 entry.

AN/FPS-100

This radar was another modernization of the Bendix AN/FPS-20 radar. See the AN/FPS-20 entry.

AN/FPS-107

This Westinghouse-built search radar operated in the L-band at 1250 to 1350 MHz.

SAGE System Compatible Height-finder Radars

To complement the search radars, height-finding radars were developed to detect aircraft at increasing altitudes. The AN/FPS-6 would serve as the standard model for much of the Cold War.

AN/FPS-6, 6A, 6B

The AN/FPS-6 radar was introduced into service in the late 1950s and served as the principal height-finder radar for the United States for several decades thereafter. Built by General Electric, the S-band radar radiated at a frequency of 2700 to 2900 MHz. Between 1953 and 1960, 450 units of the AN/FPS-6 and the mobile AN/MPS-14 version were produced.

AN/FPS-26

Avco Corporation built this height-finder radar that operated at a frequency of 5400 to 5900 MHz. This radar deployed in the 1960s.

AN/FPS-89

General Electric produced this improved version of the AN/FPS-6 height-finder radar in the early 1970s. Operating in the S-band, this high-power radar was capable of detecting targets at a range of over 110 miles.

AN/FPS-90

Martin Marietta produced this high-powered version of the AN/FPS-6 height-finder radar. See the AN/FPS-6 entry.

AN/FPS-6

AN/FPS-116

This radar was another modernized version of the AN/FPS-6 height-finder radar. See the AN/FPS-6 entry.

Gap-Filler Radars

Gap-filler radars were designed to cover areas where enemy aircraft could fly low enough to evade detection by distant long-range search radars. Between 1957 and 1962, some 200 AN/FPS-14 and AN/FPS-18 models were built.

AN/FPS-14

This medium-range search radar was designed and built by Bendix as a SAGE system gap-filler radar to provide low-altitude coverage. Operating in the S-band at a frequency between 2700 and 2900 MHz, the AN/FPS-14 could detect at a range of 65 miles. The system was deployed in the late 1950s and 1960s.

AN/FPS-18

This medium-range search radar was designed and built by Bendix as a SAGE system gap-filler to provide low-altitude coverage. The radar operated in the S-band at a frequency between 2700 and 2900 MHz. The system deployed in the late 1950s and 1960s.

AN/FPS-19

This Raytheon gap-filler radar was deployed on the Distant Early Warning (DEW) Line. It operated in the S-band.

AN/FPS-19

North Warning System Radars

The North Warning System replaced the DEW Line system in the late 1970s. New equipment came with the change in system designation. A key component of the modernization was a long-range radar system formally known as Seek Igloo. The system is based around the AN/FPS-117.

AN/FPS-117

This 3-D long-range radar was built by GE Aerospace for use at Alaskan sites and on the Northern Warning System. The radar operated at 1215 to 1400 MHz and had a range of about 220 miles.

AN/FPS-124

This medium-range radar was built by Unisys to serve as an unmanned gap-filler radar on the North Warning System.

Ballistic Missile Early Warning System (BMEWS) Radars

With the advent of ballistic missiles, millions of dollars were spent to research, develop, test, and deploy BMEWS radars.

AN/FSS-7

This radar was a modified AN/FPS-26 height-finder radar produced by Avco Corporation to detect submarine-launched ballistic missiles. The system deployed at seven sites in the 1970s. Six sites were phased out during the early 1980s. The remaining unit continued in operation in the southeast for a few more years to provide coverage over Cuba.

AN/FPS-17

AN/FPS-17

With the Soviet Union apparently making rapid progress in its rocket program, in 1954 the United States began a program to develop a tracking radar. General Electric was the contractor and Lincoln Laboratory was the subcontractor. This tracking radar, the AN/FPS-17, was conceived, designed, built, and installed for operation in less than two years. Installed at Laredo AFB in Texas, the first AN/FPS-17 was used to track rockets launched from White Sands, New Mexico. The radar was unique; it featured a fixed-fence antenna that stood 175 feet high and 110 feet wide. The transmitter sent out a VHF pulse at a frequency between 180 to 220 MHz. Units were installed in the late 1950s at Shemya Island in the Aleutians and in Turkey. The unit at Shemya subsequently was replaced by the Cobra Dane (AN/FPS-100) radar.

AN/FPS-49, 49A

This large radar was built by RCA for use in the BMEWS program and the satellite-tracking program that deployed in the 1960s. The prototype unit operated at Moorestown, New Jersey. Two additional units were installed in Greenland and England. The radar frequency operated in the Ultra High Frequency (UHF) band and could track objects beyond 3,000 miles.

AN/FPS-50

This was a BMEWS program surveillance radar that used a large, fixed-antenna fence system. Two beams were projected from the antenna array. Objects passing through the lower-angled beam provided initial data and warning for the North American Air Defense Command (NORAD). Data produced when the object passed through the upper beam allowed computation of trajectories on launch and target points. The radar operated in the UHF range at 425 MHz. General Electric, Heavy Military Electronics Department, installed these systems at Clear, Alaska, and Thule, Greenland, during the early 1960s.

AN/FPS-85

This UHF, 3-D, phased-array radar was designed by Bendix for satellite tracking. Built in the early 1960s at Eglin AFB in Florida, it was the first phased-array unit in the United States. A fire destroyed the first model in 1965. A rebuilt model became operational in 1969. The southward-sloped structure contained a square transmitter face placed alongside a larger octangular receiving face. The transmitter operated at a UHF frequency of 442 MHz. The AN/FPS-85 was also used to detect submarine-launched ballistic missiles.

AN/FPS-92

This improved version of the AN/FPS-49 tracking radar was used in the BMEWS Program. Built by RCA, this radar was installed at Clear, Alaska, in the late 1960s. The radar operated in the UHF band around 425 MHz and had a range of over 3,000 miles.

AN/FPS-108

AN/FPS-108 (Cobra Dane)

Cobra Dane was a large single-faced, phased-array radar built by Raytheon in the 1970s on Shemya Island in the Aleutians. As the main component of the Cobra system, the radar had the primary role of providing intelligence on Soviet test missiles fired at the Kamchatka peninsula from locations in southwestern Russia. Other components of the Cobra system included the ship-based Cobra Judy phased-array radar and the aircraft-based Cobra Ball and Cobra Eye radars. In addition to determining Soviet missile capabilities, Cobra Dane had the dual secondary role of tracking space objects and providing ballistic missile early warning. The radar antenna face of the building measured about ninety feet in diameter and contained some 16,000 elements. The L-band radar had a range of 2,000 miles and could track space objects as far as 25,000 miles away.

AN/FPS-115

Raytheon built the PAVE PAWS phased-array, missile-warning radar deployed during the early 1980s. At the four continental United States sites, the ninety foot diameter circular panel radars were mounted on two walls of a triangular-shaped pyramid structure. The antenna operated at a frequency of 420 to 450 MHz. PAVE PAWS could detect targets at ranges approaching 3,000 miles.

AN/FPS-118 (OTH-B)

Designed and built by GE Aerospace, the OTH-B radar was deployed on the east and west coasts in the 1980s. The system reflected the radar beam off the ionosphere to

Radar System Classification Methods

AN/FPS-115 (Front, Back)

detect objects from ranges of 500 to nearly 2,000 miles. The transmitter arrays operated at frequencies between 5 and 28 MHz. Fixed transmitter and receiving antenna arrays were separated by a distance of 80 to 120 miles.

PARCS

The acronym, PARCS, stands for Perimeter Acquisition Radar attack Characterization System. This huge structure was built as the main sensor for the Army's Safeguard missile system that deployed north of Grand Forks, North Dakota. Upon shutdown of Safeguard in 1976, the Air Force took over the huge UHF phased-array radar for use in tracking ballistic missiles and objects in space.

Federal Aviation Administration (FAA) Radars

Beginning in the late 1950s, the Civil Air Administration (predecessor to the FAA) and the DoD began to cooperate to reduce duplication. By the late 1980s most radars performing air search for the military were operated by the FAA in the joint surveillance program. Because it is a civilian agency, the FAA uses a different radar designation system.

ARSR-1

This Raytheon-built Air Route Surveillance Radar (ARSR) was used by the FAA Authority Radar beginning in 1958. It operated on a L-band frequency of 1280 to 1350 MHz with a maximum range of 200 miles.

ARSR-2

Developed by Raytheon in the 1960s as a replacement for the ARSR-1, this radar also operated in the L-band and had a similar maximum range to the ARSR-1.

ARSR-3, 3D

This Westinghouse-built search radar was used by the FAA in the Joint Surveillance System (JSS). The radar operated in the L-band at 1250 to 1350 MHz and detected targets at a distance beyond 240 miles. The D model had height-finder capability.

ARSR-4

The FAA began installing this Westinghouse-built 3-D air surveillance radar in the 1990s for the JSS system. By the late 1990s this radar will have replaced most of the 1960s-vintage AN/FPS-20 variant search radars.

Command and Control Systems

Semi-Automatic Ground Environment (SAGE) System

The SAGE system was conceived by the Lincoln Laboratory at MIT in the early 1950s to receive various sensor inputs and to detect, identify, track, and provide interceptor direction against air-breathing threats to North America. The SAGE system removed Ground Control Intercept functions from several of the radar sites and reduced manpower requirements. The first SAGE control center became operational in 1958 and the system was completed in 1961. The number of SAGE centers was reduced from about two dozen in 1962 to six in 1969. These remaining six were retired in 1983. The SAGE system featured the IBM AN/FSQ-7 (Whirlwind II) large-scale, vacuum-tube, electronic, digital computer.

Backup Interceptor Control (BUIC) System

Because the SAGE system was vulnerable to attack from Soviet intercontinental ballistic missiles (ICBMs), the Air Force sought an alternative command and control system. In the early 1960s, some radar sites increased manning to pre-SAGE levels and manually assumed pre-SAGE Ground Control Intercept functions. The sites given this ability to perform command and control functions were called BUIC I sites. Starting in 1965, BUIC II sites became operational. BUIC II sites featured the Burroughs AN/GSA-51 computer that allowed the automatic processing of data from various radar sites. BUIC III sites became operational in the late 1960s. These sites hosted the more capable Burroughs D825 digital computer and could support operations at eleven control consoles. During the early 1970s two BUIC sites were designated to serve as backup to each of the remaining six SAGE centers. Most BUIC sites were removed from service in the mid-1970s. The BUIC center at Tyndall AFB, Florida, remained in service until the early 1980s.

Joint Surveillance System (JSS)

JSS was an Air Force/FAA cooperative effort to provide a peacetime air surveillance and control system to replace SAGE and BUIC systems. Region Operations Control Centers (ROCCs) opened in 1983 and featured the H5118ME computer.

References

Information on the various systems was obtained from many sources. Significant sources included *U.S. Radar Survey Section 3—Ground Radar Change 1* (Washington, DC: National Defense Research Committee under the authority of Joint Communications Board of the Joint Chiefs of Staff, June, 1945); Eli Brookner, *Radar Technology* (Boston: Artech House, 1977); Bernard Blake, ed., *Jane's Radar and Electronic Warfare Systems, 1994–1995,* 6th ed. (Alexandria, VA: Jane's Information Group, Inc. 1994).

Part III

Site Listings

Radar Site Annotations

The radar sites are listed by state. Codes in the heading for each site are as follows:

- L stands for Lashup
- LP is a Lashup/permanent network hybrid
- P is one of the original permanent stations
- M is a first phase mobile station
- RP is a replacement of a permanent station
- SM is a second phase mobile station
- TM is a third phase mobile station
- Z is the designation of a station added to the network after the initial deployment and the designation of **all** stations after July 31, 1963.

One Air National Guard site was incorporated into the system in the 1960s. It was given the code "CW." An Army Missile Master (MM) site also was incorporated in the 1960s.

In the 1980s, most radars were turned over to the FAA for operation under the Joint Surveillance System. The radars having a "Z" designation that remained in operation were redesignated as "J" stations.

After the alpha-numeric designation, the entry heading lists the location of the site. Often sites are referenced by more than one name. All referred names are listed side-by-side. The codes in parentheses following the site location refer to entries in the Appendix, which provide additional details on each site's deployment program. The following is an example of the codes, and how to interpret them.

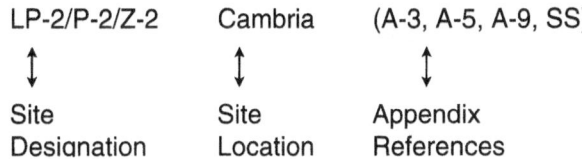

LP-2/P-2/Z-2	Cambria	(A-3, A-5, A-9, SS)
↕	↕	↕
Site Designation	Site Location	Appendix References

Alabama

TM-196/Z-196/Z-249 — Dauphin Island (A-8, A-9/SS)

This site and the attendant 693rd AC&W Squadron became operational in January 1959. The site has AN/FPS-20 search radar and two AN/FPS-6A height-finder radars. Dauphin Island began operating an AN/FPS-7 radar in 1960 and joined the SAGE system. In 1961 this site was an ADC/FAA joint-use facility. However, it was removed from joint-use status in 1962 and then put back in joint-use status in 1963. A second radar squadron, the 635th, arrived from Fort Lawton, Washington, in January 1963. Two years later, the site was again removed from joint-use status. The Air Force deactivated the 693rd in September 1970. The 635th was deactivated on July 1, 1974. In the early 1970s this unit operated AN/FPS-93 and AN/FPS-6 radars.

TM-197/Z-197 — Thomasville (A-8)

This site became operational in late 1959 when a test model of the AN/FPS-35 was installed for evaluation. This unit became permanent in the 1960s. The assigned radar squadron, the 698th, was deactivated in December 1969.

TM-199/Z-199 — Eufaula (A-8)

This site became operational in early 1959 when a test model of the AN/FPS-24 was installed for evaluation. In the 1960s the installation became permanent and part of the SAGE system. The tending 609th Radar Squadron was deactivated in September 1968.

Z-336/J-12 — Grand Bay

This FAA radar site fed into the JSS system that operated in the 1980s and 1990s.

SAGE — Gunter AFS

Activated on September 8, 1957, this Sector Director Center controlled operations within the Montgomery Air Defense Sector. The SAGE center was deactivated in 1966.

ALASKA

Alaskan radar sites are thoroughly detailed in D. Colt Denfield's, *The Cold War In Alaska: A Management Plan for Cultural Resources, 1994–1999* (Alaska District, U.S. Army Corps of Engineers, 1994). The identities of different sites are listed below.

Mobile Radar Sites

From 1950 through 1953 mobile radar units were operating at the following locations: Bethel AFS, Farewell Airport, Gambell on St. Lawrence Island, Ladd AFB, and Middleton Island AFS.

Permanent AC&W Sites

Permanent Aircraft Control and Warning Stations operated at the locations listed below. Parentheses indicate approximate operational dates. Sites with no end dates are part of the North Warning System Long-Range Radar

An Alaskan radar station, circa 1954. (Official U.S. Air Force photo courtesy National Archives.)

System (LRRS) formally known as Seek Igloo. Using a Minimally Attended Radar (MAR) featuring the AN/FPS-117, some of these sites have undergone considerable reconstruction. Four of these sites (marked with *) received unique geodesic domes in the early 1980s.

Bethel AFS (1958–1963), Campion AFS (1952–1985), Cape Lisburne AFS (1953–), Cape Newenham AFS (1954–), Cape Romanzof AFS * (1953–), Fort Yukon AFS (1958–), Indian Mountain AFS * (1953–), King Salmon Airport (1951–), Kotzebue AFS (1950–), Middleton Island AFS (1958–1963), Murphy Dome (1952–), Northwest Cape on St. Lawrence Island (1953–), Ohlson Mountain AFS (1958–1963), Sparrevohn AFS * (1954–), Tatalina AFS * (1954–), Tin City AFS (1953–), Unalakleet AFS (1958–1963). A site at Fire Island AFS (1951–early 1980s) also served as Regional Combat Command Center during the 1960s.

> ## THE AIR FORCE STATION AT SPARREVOHN
>
> Sparrevohn AFS typified many Air Force Radar Stations in Alaska. Located in the remote wilderness about 200 miles west of Anchorage, Sparrevohn's nearest neighbor was Lime Village located eighteen miles to the northeast. However, with the local terrain, Lime Village must have seemed to be eighteen hundred miles away. Airmen with the 719th Aircraft Control and Warning (AC&W) Squadron observed that Sparrevohn was a small, highly developed island of civilization in the middle of nowhere. The 719th began operating the radar in March 1954. Additional personnel arrived in 1957 to operate the White Alice Communications System composed of a communications center, power plant, and four billboard antennas. With the replacement of White Alice in 1978, eighty personnel departed. In the mid-1980's, the Air Force installed a Minimal Attended Radar (MAR) that could be operated by six contract personnel.
>
> Construction had begun in 1959 when the Air Force parachuted equipment and personnel to clear land for a primitive airstrip. This airstrip became the station's only access to the outside world. While the Gaasland Company had the contract for the AC&W station, the Northern Corporation put in the White Alice facility. Both projects proved challenging as the radar dome and White Alice site were located on top of a mountain 1,500 feet higher than the base camp. To connect the two sites, a six-mile zigzagged road featuring eleven switchbacks was constructed. Because severe storms could make the road impassable, a tramway was built to ensure that men operating the radar and communication stations would have enough food and supplies to survive the winter.
>
> The base camp consisted of wooden structures including a headquarters building, administration building, airmen quarters, Bachelor Officers Quarters, communications center, power plant, water storage building, warehouse, and recreation center and gymnasium. Most of the inhabited structures were connected by covered walkways. Nestled in the wilderness among rivers and lakes, fishing and hunting opportunities were plentiful.
>
> Source: Denfield, pp. 28–29, 165.

DEW Line/North Warning System

Construction of the Distant Early Warning (DEW) Line began in 1953. An extension covering the Aleutians was authorized in 1957. By 1960 the Alaska portion of the line had three main stations, ten auxiliary, and ten intermediate stations. Only the main and auxiliary sites had rotating radars. Intermediate sites had gap-filling, doppler-type radar fences and operated until 1963. The Aleutian extension sites were closed in 1969. In the 1980s the DEW Line was phased out and replaced with the North Warning System featuring long-range AN/FPS-117 and unmanned short-range AN/FPS-124 radars.

Main Sites: Barter Island (1957–), Point Barrow (1957–), Cold Bay (1959–). These sites are now LRRS facilities.

Auxiliary Sites: Bullen Point/Flaxman Island (1957–1971), Oliktok Point (1956–), Point Lay (1956–), Point Lonely (1954–), Wainwright (1954–), Cape Sarichef (1959–1969), Driftwood Bay (1959-1969), Nikolski (1959–1969), Port Heiden (1959–1969), Port Moller (1959–1969).

Intermediate Sites: Brownlow Point (1957–1963), Camp Sabine (1957–1963), Camp Simpson (1957–1963), Collison Point (1957–1963), Demarcation Point (1953–1963), Icy Cape (1957–1963), Kogru River (1957–1963), McIntyre (1957–1963), Nuvagapak (1953–1963), Peard Bay (1957–1963).

Ballistic Missile Early Warning System (BMEWS) Sites

One of three BMEWS sites was located at Clear. The detection radar, known as an AN/FPS-50, used a fixed antenna system that was constructed in 1959 and 1960. Operation of this system began in 1961. In 1966 an AN/FPS-92 tracking radar was added to the site.

Over-the-Horizon Backscatter (OTH-B) Site

The third of a system of OTH-B radar sites was scheduled for completion in the early 1990s. Substantial progress was made on the receiving station located north of Gakona before the system was canceled. Consequently, the Air Force used the site for its High Frequency Active Auroral Research, an experimental communications program.

Cobra Dane Site

Located at Eareckson AFS on Shemya Island in the Aleutians, Cobra Dane was an AN/FPS-108 giant phased-array radar designed primarily to track Soviet missile tests. Completed in 1976, the site became operational in 1977.

Arkansas

M-91/Z-91 — Texarkana (A-6, A-9/GCI)

In 1955 the 703rd AC&W Squadron began operations at Texarkana. By 1958 the 703rd was using AN/FPS-20 and AN/MPS-14 sets. In 1960 this site was also performing air traffic control duties for the FAA. By 1966, the search radar was upgraded to an AN/FPS-91A set. The Air Force deactivated the 703rd in September 1968.

SM-143 — Walnut Ridge (A-7)

This site became operational in 1956. The 725th AC&W Squadron operated an AN/MPS-11 set here. In 1958 the AN/FPS-6 replaced an AN/MPS-8 height-finder radar that had been installed a year earlier. In March 1963 the Air Force ordered this site to close. Operations ceased two months later.

ARIZONA

M-92/Z-92 — Mt. Lemmon (A-6, A-9/SS)

The 684th AC&W Squadron began operations using AN/MPS-7, AN/MPS-14, and AN/TPS-10D sets at Mt. Lemmon in August 1956. The AN/TPS-10D was soon retired. By 1959 an AN/FPS-20 had replaced the AN/MPS-7 search radar. In 1961 M-92 became a SAGE center. During the following year, an AN/FPS-6 replaced the AN/MPS-14 height-finder radar and the AN/FPS-20 was upgraded to become an AN/FPS-67. The 684th was deactivated in December 1969.

M-93 — Winslow (A-6, A-9/SS)

Winslow became operational in 1956 with the 904th AC&W Squadron using an AN/MPS-11 radar. In 1958 an AN/FPS-6A height-finder radar was added. In March 1963 the Air Force ordered Winslow closed; operations ceased on April 30.

M-128 — Kingman (A-6)

This site, operational in 1955, was placed in standby status in 1957 and deactivated in 1958. During Kingman's brief lifespan, the 659th AC&W Squadron operated AN/MPS-7 and AN/MPS-14 sets.

SM-162 — Vincent AFB, Yuma (A-7, A-9/SS)

The 864th AC&W Squadron began operations in 1956. In March 1963 the Air Force ordered the site and its AN/MPS-7 and AN/MPS-14 radars to shut down. Operations ceased on April 30.

TM-181/Z-181 — Luke-Williams Range/Ajo (A-9/NCC)

Construction costs came to approximately $7.4 million for 100 structures located within housing, cantonment, operations, ground-to-air transmitter receiver (GATR) areas. This site became active in 1959. At that time, the 612th AC&W Squadron operated AN/FPS-20A and AN/FPS-6 sets. In 1961 Luke Range became a SAGE center. By 1963 an AN/FPS-7C had assumed search duties and height-finder radar chores were being performed by AN/FPS-6A and AN/FPS-26 radars. The 612th was deactivated on December 31, 1969.

Housing units were moved to Gila Bend and the remaining buildings were abandoned. A minimal Air Force and FAA presence has been kept to operate some instrumentation and radio signal relay equipment.

Z-247/J-29 — Phoenix/Humboldt Mountain/Cave Creek

A detachment of the Luke-based 4629th Air Defense Squadron came to this FAA-operated site in late 1972 to set up and operate an AN/FPS-6 height-finder radar in conjunction with the FAA-operated ARSR-1 set.

SAGE — Luke AFB

Air Defense Command (ADC) activated a direction center at Luke AFB on June 15, 1959. This center controlled the Phoenix Air Defense Sector until disestablishment in 1966.

California

L-38 — Half Moon Bay (A-1)

This site, near San Francisco, was established in 1947 by the 505th AC&W Group for the purposes of ground-control interception training. The site used a World War II vintage AN/CPS-5 radar. The site was incorporated into the Lashup system using an AN/CPS-6. In October 1951 Mt. Tamalpais (P-38) assumed radar coverage for this area.

L-39 — Minter Field (A-1)

In late 1950 this Lashup site was operating an AN/CPS-4 radar. In June 1952 Mt. Laguna (P-76) assumed coverage for this area.

L-40 — Edwards AFB (A-1)

In October 1950 an AN/CPS-5 was operational at Edwards. Site P-59 at Atolia assumed area coverage in February 1952.

L-41 — Camp Cooke (A-1)

In October 1950 an AN/TPS-1C radar operated from this coastal site. Site P-15 at Santa Rosa Island assumed area coverage in April 1952.

L-42 — Point Hueneme (A-1)

Port Hueneme operated an AN/TPS-1C radar in late 1950. Santa Rosa Island assumed coverage for this site in April 1952.

L-43 — Fort MacArthur (A-1)

In December 1950 an AN/TPS-1B radar operated at this coastal site near Los Angeles. Coverage was assumed in January 1952 by site P-74 at Madera.

LP-2/P-2/Z-2 — Cambria (A-3, A-5, A-9/SS)

The 775th AC&W Squadron began operating AN/FPS-3 search and AN/CPS-4 height-finder radars at this site in January 1952. In 1955 the height-finder radar was replaced by an AN/FPS-6. An AN/FPS-8 medium-range search radar also operated for a short time from 1955 to 1956. In 1959 the site featured two AN/FPS-3, and AN/FPS-6 and AN/FPS-6A radars. In 1961 the site received an AN/FPS-7 radar featuring ECCM capability. In 1961 the site also replaced the AN/FPS-6A height-finder radar with an AN/FPS-6B model. In April of that year, Cambria was incorporated into the SAGE system. In 1963 an AN/FPS-26A replaced the AN/FPS-6B height-finder radar. The facility came under Tactical Air Command (TAC) jurisdiction in 1979.

P-15 — Santa Rosa Island (A-3)

Activation of this site allowed for the shutdown of Lashup sites at Camp Cooke (L-41) and Point Hueneme (L-42). By 1952 the 669th AC&W Squadron operated two AN/FPS-10 radars at this island location. In 1955 the Air Force added an AN/FPS-3 search radar to the facility. During the following year, this was replaced with an AN/GPS-3. 1958 saw the addition of an AN/MPS-14 long-range height-finder radar. Operations ceased on March 31, 1963, and the 669th moved to Lompoc (RP-15).

RP-15/Z-15 — Lompoc/Vandenberg

Activated in 1963 by the 669th Radar Squadron (SAGE), Lompoc AFS initially hosted AN/FPS-67 search and AN/FPS-6A height-finder radars. The 669th was deactivated in June 1968.

LP-33/P-33/Z-33/J-83 — Klamath/Cresent City (A-4, A-5, A-9/SS)

Beginning in April 1952, the 777th AC&W Squadron began operating AN/FPS-3 and AN/FPS-4 radars at this northern California coastal site. In 1956 the Air Force added an AN/GPS-3 to the facility. By 1958 an AN/FPS-20 and AN/FPS-6 had replaced the initial pair of radars. During the following year, an AN/FPS-6A height-finder radar replaced the AN/GPS-3. During 1960 the 777th became a SAGE radar squadron. This meant the radar data was now being fed to a Region Command and Control Center. By the end of 1961, the AN/FPS-20A had been upgraded and redesignated as an AN/FPS-66. By 1966 there was an AN/FPS-27 long-range, height-finder radar in operation there. The site came under TAC jurisdiction beginning in 1979. In the 1980s much property was turned over to the National Park Service. The operations area became a FAA/USAF joint-use facility. In 1995, the FAA operated an AN/FPS-66A search set.

LP-37/P-37/Z-37/J-34 — Peak Hill Road/Point Arena (A-3, A-5, A-9/NCC)

By December 1951 the 776th AC&W Squadron was operating AN/FPS-3 and AN/FPS-4 radars at this site located near the coast 120 miles north of San Francisco. In

1955 the 776th received an AN/FPS-8 that subsequently was converted to an AN/GPS-3. In 1958 AN/FPS-20 and AN/FPS-6 radars had replaced the original sets. An AN/FPS-6B joined the site in 1960, the year the site came into the SAGE system. Point Arena AFS replaced its AN/FPS-20 with an AN/FPS-24 radar in 1961. By 1963 the 776th Radar Squadron (SAGE) had replaced its AN/FPS-6 height-finder radars with AN/FPS-26A and AN/FPS-90 models. The 776th held additional responsibilities during the 1960s as Point Arena was designated as a Backup Intercept Control site for both the BUIC I and BUIC II programs. In 1979 the site came under TAC jurisdiction. The 776th subsequently was deactivated and an element of the 26th Air Defense Squadron continued operations. A reorganization in 1987 placed the site under the Southwest Air Defense Sector of the 25th Air Division.

P-38/Z-38/J-33 — Mt. Tamalpais/Mill Valley (A-2)

The 666th AC&W Squadron began operating a pair of AN/CPS-6B radars at this Bay-area site in late 1951. In 1955 the site received an AN/FPS-8 that subsequently was converted to an AN/GPS-3. During 1956 an AN/FPS-4 height-finder radar operated here. In 1958 the AN/FPS-4 was superseded by an AN/FPS-6 set. In late 1960 this site began feeding data into the SAGE system. This site began operating an AN/FPS-7 frequency-diversity (FD) radar in 1960. By 1961 the 666th operated this radar along with the AN/FPS-6 and AN/FPS-6B height-finder radars. Mill Valley came under TAC jurisdiction in October 1979. During the 1980s, most of the property was turned over to the National Park Service and the FAA. The Air Force retained control of the height-finder tower. In 1995 the FAA operated an AN/FPS-66A search set.

P-39 — San Clemente Island (A-3)

The 670th AC&W Squadron began operations in May 1952 with a single AN/FPS-3 radar. A year later, an AN/FPS-4 height-finder radar joined the site. In 1955 an AN/FPS-8 came to the island. This radar subsequently was converted to a AN/GPS-3. In 1956 an AN/FPS-6 height-finder radar replaced the AN/FPS-4. This site was deactivated in 1960 and relocated to San Pedro Hill (RP-39).

RP-39/Z-39/J-31 — San Pedro Hill/Fort MacArthur (A-9/SS)

Assuming coverage duties from the closed San Clemente Site (P-39), this Los Angeles area site was an ADC/FAA joint-use facility that began operations in 1961 with an FAA ARSR-1C radar. The Air Force provided the height-finder radars. In 1963 these radars included AN/FPS-6B and AN/FPS-27 sets. In 1966 an AN/FPS-27 radar was in operation. The tending 670th Radar Squadron was deactivated in April 1976. The FAA assumed control of the property although the Air Force continued to operate the height-finder tower.

L-37/P-58/Z-58 — Mather AFB (A-1, A-3)

In 1950 the site operated an AN/CPS-6 as a Lashup site. This radar was incorporated into the permanent network in 1951. The site was operated by the 668th AC&W Squadron. The squadron operated a pair of AN/CPS-6B radars throughout the decade until a conversion to AN/FPS-20 and AN/FPS-6 and 6B radars was initiated. By 1960 this site was a joint-use facility, handling air traffic control for the FAA. During the following year, the site began providing data for the SAGE system. Also in 1961 the height-finder radars were removed. In September 1961 the 668th was deactivated and the radar came under FAA control.

P-59/Z-59 — Atolia, Boron (A-3, A-9/NCC)

In February 1952 the 750th AC&W Squadron assumed coverage responsibilities formerly held by the Edwards AFB site (L-40) and was operating two AN/FPS-10 radars at this new site. The AN/FPS-10 search radar remained until 1959. In 1958 an AN/FPS-6 replaced the AN/FPS-10 height-finder radar. A second height-finder radar (AN/FPS-6A) was installed in 1959. In 1961 the facility provided data for the regional SAGE center and became an operational ADC/FAA joint-use radar. By this time the AN/FPS-10 had been replaced by an AN/FPS-20 search radar. However, this radar was soon replaced by an AN/FPS-35 FD radar. By 1963 this radar operated with AN/FPS-26A and AN/FPS-90 height-finder radars. The 750th was deactivated in June 1975.

LP-74/P-74/Z-74 — Madera (A-4, A-5)

Assuming the radar coverage of a Lashup site at Fort MacArthur (L-43), the 774th AC&W Squadron began operating AN/FPS-3 and AN/FPS-4 radars from this location in January 1952. In 1958 the height-finder radar was replaced by AN/FPS-6 and AN/FPS-6A radars. In 1959 an AN/FPS-20 search radar superseded the AN/FPS-3. In 1960 the Air Force upgraded an AN/FPS-6A to become a 6B and integrated the site into the SAGE system. The Air Force then upgraded the search radar; first to an AN/FPS-20A and then to an AN/FPS-66. By 1963 this AN/FPS-66 search radar operated in conjunction with AN/FPS-6 and AN/FPS-90 height-finder radars. This site was closed on April 1, 1966.

P-76/Z-76/J-30 — Mt. Laguna (A-3, A-9/GCI)

Operations at this peak began in April 1952. Within two months the radar assumed coverage formerly provided by the Minter Field Lashup site (L-34). At that time the 751st AC&W Squadron operated AN/CPS-4 and AN/FPS-3 radars. An AN/FPS-8 replaced the AN/CPS-4 in 1955. This radar then was converted to an AN/GPS-3 in 1956 and removed in 1960. 1956 also saw the arrival of an AN/FPS-6 at the site. The site became integrated into the SAGE system in 1961. By 1962 the 751st operated an AN/FPS-7 FD search radar and AN/FPS-6 and 6B height-finder radars. In 1963 the 6B was upgraded to an AN/FPS-90 set. Mt. Laguna became a joint-use ADC/FAA facility around 1965. In 1979 the facility

California

came under TAC jurisdiction. In the 1980s the FAA assumed greater control, leaving the Air Force only responsible for the height-finder tower.

M-96/Z-96 — Almaden (A-6, A-9/GCI)

Great difficulties were encountered in construction at this site. During fiscal year 1957, the Air Defense Command held up construction due to funding shortfalls. The radars finally came on line with AN/FPS-20 and AN/MPS-14 sets in 1958. A year later, an AN/FPS-6A height-finder radar joined the site. In 1961 the site received an AN/FPS-24 radar, but could test the radar only on a not-to-interfere basis with television transmissions. This radar became operational in 1962. In 1963 an AN/FPS-90 performed the height-finder duties. The 682nd AC&W Squadron operated this site. The site came under TAC jurisdiction in 1979.

SM-157/Z-157 — Red Bluff (A-7)

This site became operational in 1956 under the jurisdiction of the 859th AC&W Squadron. The site used AN/MPS-8 and AN/MPS-11 radars. The AN/MPS-11 remained until 1963. In 1959 the Air Force placed AN/FPS-6 and 6A height-finder radars and removed the AN/MPS-8 from Red Bluff. SAGE operations began in 1960. In 1964 Red Bluff became an FAA/ADC joint-use facility, using the AN/FPS-67 search and AN/FPS-6 and AN/FPS-90 height-finder radars located on site. The 859th was deactivated in September 1970.

Z-342/J-32 — Pasa Robles AFS

The FAA began operations in 1960 at this site located twenty-five miles east of San Luis Obispo. The Air Force placed a height-finder radar at the site in 1980. Also in 1980 the facility became a JSS site.

Over-the-Horizon-Backscatter (OTH-B) — Tule Lake

During the late 1980s, the Air Force constructed a receiver for the west coast OTH-B radar system at this site near Alturas. The Air Force began operations in late 1990. However, the site was placed on a standby status in 1991.

PAVE PAWS — Beale AFB

The PAVE PAWS radar site at Beale became operational in 1980. A large three-sided structure, the PAVE PAWS hosted two large AN/FPS-115 phased-array radar antennas. The main mission of PAVE PAWS was to warn of submarine-launched ballistic missiles.

SAGE — Beale AFB

Activated on February 15, 1959, this direction center for the San Francisco Defense Sector remained active until December 1963.

SAGE — Norton AFB

Also activated on February 15, 1959, this Los Angeles Defense Sector direction center remained active until June 1966.

Colorado

Z-212 — Denver
This ARSR-1 joint-use radar began feeding information into the Air Defense Command network in 1963.

Z-215 — Grand Junction
This FAA ARSR-2 joint-use radar began feeding information into the Air Defense Command network in 1963.

Z-222 — Trinidad
This ARSR-2 joint-use radar began feeding information into the Air Defense Command network in 1963.

CW-59 — Buckley Field
This Air National Guard radar began twenty-four-hour-a-day operations in 1960 with an AN/FPS-8 radar. This radar fed information into the Air Defense Command network during the 1960s.

Z-239 — Greeley
By the mid-1960s the Air National Guard was supporting the Air Defense Command with its AN/FPS-8 from this location.

NORAD Combat Operations Center, Cheyenne Mountain
The development of this site is discussed in Part I.

Florida

M-114/Z-114 — Jacksonville NAS (A-6, A-9/GCI)

This was the last of the first phase of mobile radars to be activated. This site came on line in 1959 with the 679th AC&W Squadron operating AN/FPS-3, AN/FPS-8, and AN/MPS-14 radars. In 1962 the site was relocated within the Jacksonville Naval Air Station and reestablished featuring an AN/FPS-66 radar and a pair of AN/FPS-6 height-finder radars. By 1963 it was an FAA/ADC joint-use site. The site came under TAC jurisdiction in 1979. In 1981 the Air Force deactivated the 679th.

M-129/Z-129 — MacDill AFB (A-6)

This site became the first operational mobile radar under the first phase of the program. Activated by the 660th AC&W Squadron on December 6, 1954, this site initially used an AN/MPS-7 radar. By 1958 this site also had AN/GPS-3 and AN/MPS-14 radars. During the following year the AN/GPS-3 and AN/MPS-7 sets were replaced by AN/FPS-20A search and AN/FPS-6B height-finder sets. In 1960 MacDill became a SAGE system contributor. In 1961 an AN/FPS-7B assumed search duties and an additional height-finder radar was added in the form of an AN/FPS-26. In 1963 an AN/FPS-90 height-finder radar replaced the AN/FPS-6B. Around 1966 this became an FAA/ADC joint-use site featuring an AN/FPS-7E radar. In 1979 this site came under TAC jurisdiction.

TM-198/Z-198/J-11 — Tyndall AFB (A-8, A-9/NCC)

In 1957 this station was the first station operating for the third phase of the mobile radar program. Activated by the 678th AC&W Squadron, this station became operational to support the BOMARC surface-to-air missile program. In 1958 the site was operating with an AN/FPS-20 search radar and a pair of AN/FPS-6 height-finder sets. During 1960 Tyndall joined the SAGE system. In 1962 the search radar was upgraded and redesignated as an AN/FPS-64. Around 1965 Tyndall became an FAA/ADC joint-use facility. It also received a BUIC II, and later BUIC III, capability to perform command and control functions. Tyndall retained this function until the 1980s. On March 1, 1970, the 678th was redesignated as the 678th Air Defense Group. On October 1, 1979, this site came under TAC jurisdiction. In 1995 an AN/FPS-64A was performing search duties.

TM-200/Z-200/J-10 — Cross City (A-8, A-9/SS)

In 1959 the 691st AC&W Squadron activated an AN/FPS-20A search and a pair of AN/FPS-6A height-finder sets at this Northern Florida site. In 1960 Cross City joined the

SAGE system. In 1962 the search radar was upgraded to an AN/FPS-66 radar. The Air Force deactivated the 691st in September 1970. The site was turned over to the FAA with the Air Force retaining the height-finder radar.

Z-209/J-07 — Key West NAS (A-9/NCC)

This Navy AN/FPS-37 radar was added to the Air Defense Command network in 1962 at the time of the Cuban missile crisis. In June 1962 the 671st Radar Squadron (SAGE) was activated here. By 1966 an AN/FPS-67B was performing the sky search duties and would remain in operation for the next thirty years. In 1979 the Key West site became a TAC facility.

Z-210/J-06 — Richmond NAS (A-9/GCI)

Beginning in 1959 an FAA ARSR-1 radar located at this site began feeding information to the Air Defense Command system. The Air Force placed a height-finder radar at the site and the facility remains as part of the JSS.

Z-211/J-05 — Patrick AFB (A-9/SS)

In 1961 this site was operational as an FAA/ADC joint-use facility featuring an AN/FPS-66 radar. In 1962 the 645th Radar Squadron was reactivated and stayed active until April 1976. In 1995 the FAA still operated an AN/FPS-66A set at this site.

Z-330/J-09 — Fort Lonesome

This JSS station was constructed in 1980 with the installation of FAA search and Air Force height-finder radars. The site is located approximately twenty-five miles southwest of Tampa.

Z-327/J-04 — Whitehouse

This facility located west of Jacksonville is an FAA-operated JSS site built in 1979. The Air Force maintained the height-finder radar.

Z-399/J-08 — Cudjoe Key

The Army first activated this site in June 1959 to track missiles traveling over the Eglin Gulf Test Range. The Air Force assumed operations in 1960. In May 1967 the site came under the jurisdiction of the USAF Security Service for a classified mission. In September 1973 the station became the host for a balloon-borne radar surveillance system. This system was operated mostly with contract personnel.

J-17 — Cape Canaveral

This site operated as part of the JSS during the 1960s. The type of radar could not be discerned from the sources.

Eglin AFB

A large AN/FPS-85 phased-array radar was built here starting in 1962. Bendix was the primary contractor. This radar subsequently was rebuilt after a devastating fire in 1965. The replacement unit became operational in 1969 and has since played an important role in tracking orbiting objects and warning against a submarine-launched ballistic missile attack.

Georgia

M-111/Z-111 — Marietta (A-6)

This site became operational in 1956 when the 908th AC&W Squadron activated AN/MPS-8 and AN/MPS-11 radars. By 1963 these radars had been removed in favor of a pair of AN/FPS-6 height-finder radars and an FAA ARSR-1A radar. Starting in 1959, Marietta also performed air traffic control duties. In 1962 the station joined the SAGE system. The 908th was deactivated in September 1968.

M-112/Z-112 — Hunter AFB (A-6)

In 1955 the 702nd AC&W Squadron began operating AN/MPS-7, AN/TPS-10D, and AN/MPS-14 radars. From 1957 to 1958 an AN/GPS-3 also saw service at Hunter. By 1959 only AN/FPS-20A and AN/MPS-14 sets were operating here. In 1961 Hunter received an AN/FPS-26 height-finder radar. In 1962 the FPS-20A was upgraded to become an AN/FPS-67 and Hunter joined the SAGE system. The 702nd was deactivated in June 1979.

SM-165 — Flintstone (A-7)

This site became operational in 1956 when the 867th AC&W Squadron activated AN/MPS-11 and AN/TPS-10D radars. In 1960 the 867th was operating AN/FPS-6 and AN/FPS-8 radars when the Air Force deactivated Flintstone due to budgetary constraints.

PAVE PAWS — Robins AFB

The Robins PAVE PAWS site became operational in 1986. Designated as an AN/FPS-115 radar, PAVE PAWS consisted of a large three-sided building. Two sides of the building contained phased-array radar antennas. PAVE PAWS provided a warning against submarine-launched ballistic missiles.

Idaho

SM-150 — Cottonwood (A-7)

Activated by the 821st AC&W Squadron, Cottonwood became operational in 1959. The site used AN/MPS-7, AN/MPS-14, and AN/FPS-6 radars. In 1962 the 821st began operations with an AN/FPS-24 search radar and additional AN/FPS-6B radar as the AN/MPS units were retired. A bearing failure in the AN/FPS-24 antenna pedestal led to an early shutdown in December 1964 as part of fiscal year 1965 cutbacks.

Z-223 — Boise

This ARSR-2 FAA joint-use radar began feeding information into the Air Defense Command network in 1963.

Z-225 — Ashton

This ARSR-2 FAA joint-use radar began feeding information into the Air Defense Command network in 1963.

Over-the-Horizon-Backscatter (OTH-B) — Mountain Home AFB

In the late 1980s, construction began on the operations center for the west coast OTH-B radar system. The Air Force began operating the system from this site in late 1990; however, in 1991 a decision was made to place the system on a standby status.

ILLINOIS

L-49 — O'Hare International Airport (A-1)

Operational in late 1950, this Lashup site used an AN/TPS-1B radar.

RP-31/Z-31 — Arlington Heights (A-9/SS)

This radar site and attendant squadron was moved to this location from Williams Bay, Wisconsin (LP-31/P-31) and activated in 1961 with AN/FPS-20A, AN/FPS-6, and AN/FPS-6B radars. The AN/FPS-20 was upgraded to an AN/FPS-67 during the following year. Also in 1962 the site began SAGE operations. In 1963 an AN/FPS-90 had replaced the AN/FPS-6B for performing height-finding duties. The 755th Radar Squadron was deactivated in September 1969.

P-70/Z-70 — Belleville (A-4, A-9/SS)

The 798th AC&W Squadron began operations at Belleville in May 1952. The site used AN/CPS-4 and AN/FPS-3 radars. The AN/FPS-3 remained in operation until 1963. The site fed data to a SAGE center beginning in 1962. A variety of height-finder radars were used at Belleville. In 1963 two AN/FPS-6 sets stood guard. Later, during the mid-1960s, this site operated with an AN/FPS-66. The 798th was deactivated in June 1968.

P-85/Z-85 — Hanna City (A-3)

In 1952 the 791st AC&W Squadron began operations at this location using AN/FPS-3 and AN/FPS-4 sets. In 1958 this station replaced the AN/FPS-3 with an AN/FPS-20 search radar and added an AN/FPS-6A height-finder radar. During 1959 the AN/FPS-4 was replaced by a second height-finder radar (AN/FPS-6B) and Hanna City joined the SAGE system. Around 1963 the AN/FPS-20A was upgraded and redesignated as an AN/FPS-67, and an AN/FPS-90 replaced the AN/FPS-6B height-finder radar. The 791st was deactivated in June 1968.

SM-137 — Carmi (A-7)

This station became operational in 1956. Budget cuts forced closure in 1957.

INDIANA

P-53/Z-53 — Rockville (A-3, A-9/NCC)

The 782nd AC&W Squadron began operating a pair of AN/FPS-10 radars at this site in May 1952. SAGE operations began in 1959. By 1960 the original radars had been phased out and replaced by two AN/FPS-6 height-finder radars and an AN/FPS-7B search radar. Rockville was deactivated on April 1, 1966.

Iowa

P-83/Z-83 — Waverly (A-3, A-9/GCI)

In 1952 the 788th AC&W Squadron activated a pair of AN/FPS-10 radars at this site. The AN/FPS-10 search radar continued operating until replaced in the 1960s with an AN/FPS-27. In 1959 an AN/FPS-6 and 6A replaced the AN/FPS-10 height-finder radar. SAGE operations also started that year. By 1963 one of the height-finder radars was replaced by an AN/FPS-90. The Air Force deactivated the 788th in September 1969.

M-122 — Dallas Center (A-6)

Operational status was achieved in 1956 with an AN/TPS-1D radar that had been moved in from Fort Snelling, Minnesota. Budget cuts forced Dallas Center to cease operations in mid-1957 and the tending 650th AC&W Squadron was deactivated shortly thereafter.

SAGE — Sioux City

Located at the municipal airport, this SAGE direction center controlled the Sioux City Air Defense Sector. The sector was activated on January 8, 1958, and deactivated in April 1966.

Kansas

P-47/Z-47 — Hutchinson (A-4, A-9/SS)

In May 1952 the 793rd AC&W Squadron began operating a pair of AN/FPS-10 radars at this site. During 1958 an AN/FPS-3 search radar saw temporary duty and a pair of AN/FPS-6A height-finder radars were installed. In late 1959 this station was performing air traffic control duties for the FAA. At this time the site operated an AN/FPS-20 search radar. In the early 1960s this radar was upgraded and redesignated as an AN/FPS-66. By 1963, height-finding duties were being performed by AN/FPS-6A and AN/FPS-26A radars. The 793rd was deactivated in September 1968.

P-72/Z-72 — Olathe NAS (A-4, A-9/NCC)

The 738th AC&W Squadron began operations in 1952 using AN/CPS-4 and AN/FPS-3 radars. The AN/CPS-4 soon was replaced by an AN/FPS-4 set that in turn was superseded by a pair of AN/FPS-6A sets in 1958. In 1958 this site also replaced the AN/FPS-3 set with an AN/FPS-20 search radar. In late 1959 this station was also performing air traffic control duties for the FAA. Olathe was tied into the SAGE system in December 1961. In the early 1960s the search radar was upgraded and redesignated as an AN/FPS-66. In September 1968 the Air Force deactivated the 738th.

Z-226 — Garden City

This facility became an FAA/ADC joint-use site in 1964.

KENTUCKY

P-82/Z-82 — Godman/Snow Mountain (A-3)

In 1952 the 784th AC&W Squadron began operating AN/FPS-3 and AN/FPS-4 radars at this peak. In 1958 these radars were replaced by AN/FPS-20 and AN/FPS-6 sets. A second AN/FPS-6 height-finder radar was added in 1961. In 1962 the search radar was upgraded to become an AN/FPS-67. SAGE operations began in 1963. The 784th was deactivated in June 1968.

M-131 — Owingsville (A-6)

Beneficial occupancy was achieved at this Phase I Mobile Radar site in late 1954. Operational status was achieved in 1956. Budget cuts forced closure in 1957.

Louisiana

M-125 — England AFB (A-6, A-9/NCC)

This site became operational in November 1955 when the 653rd AC&W Squadron activated AN/MPS-14, AN/TPS-1D, and AN TPS-10D radars. The AN/MPS-14 continued to operate until site closure. In 1958 the site was operating an AN/FPS-20 search set. In March 1963 the Air Force ordered the site to close and operations ceased on April 23.

M-126/Z-126 — Houma NAS (A-6, A-9/GCI)

In 1955 the 657th AC&W Squadron began operations using AN/MPS-14, AN/TPS-1D, and AN/TPS-10D radars. In 1958 an AN/FPS-20 search radar was in operation along with AN/MPS-14 and AN/MPS-7 units. In 1960 an AN/FPS-6 height-finder radar was added. In 1961 Houma joined the SAGE system. In 1962 the search radar was upgraded to an AN/FPS-67 and the AN/FPS-6B height-finder radar was upgraded to an AN/FPS-90. A prototype AN/FPS-28 was placed at Houma in late 1959 for field testing. The 657th was deactivated in September 1970.

TM-194/Z-248/J-14 — Lake Charles

In 1958 the 812th AC&W Squadron activated AN/FPS-3A and AN/FPS-6 radars. In 1961 the Lake Charles site was converted into a gap-filler radar site (M-125D) and an AN/FPS-18 was installed. By 1963 this AN/FPS-18 radar was no longer in service. In 1972 the Air Force constructed five buildings at a new Lake Charles site. In the following year, TAC set up an AN/TPS-43 unit on the newly constructed search tower. Eventually, the Aerospace Defense Command assumed control of the site and placed AN/FPS-93 and AN/FPS-6 radars at the site. These radars were operated briefly by a detachment of the Houston-based 630th Radar Squadron and then by the 634th Radar Squadron. This unit subsequently was deactivated in July 1974 and activities were assumed by an element of the Southeast Air Defense Sector. The site continued operations over the next two decades with approximated twenty Air Force and civilian personnel tracking aircraft attempting to illegally enter the country.

Z-246/J-13 — Slidell

In late 1972 a detachment arrived at this FAA site from the 630th Radar Squadron to set up and operate an AN/FPS-6 height-finder radar. With the deactivation of the 630th in 1977, duties were assumed by an element of the Southeast Air Defense Sector, which was headquartered in Pearl River.

MAINE

L-1 — Dow AFB (A-1)

The Dow Lashup site became operational in June 1950 using AN/CPS-5 and AN/TPS-10A radars. Operations ceased in October 1951 as coverage was assumed by site P-65 in Charleston.

L-2 — Fort William (A-1)

This Lashup site used an AN/TPS-1B long range search radar. Tracking operations commenced in June 1950 and ceased in October 1951. However, training for Air National Guardsmen continued there afterwards.

L-50 — Limestone (A-1)

Limestone came on line in February 1951 using an AN/TPS-1B long-range search radar. The site ceased operations in February 1951 as coverage was assumed by site P-80 at Caswell.

P-13/Z-13 — Brunswick (A-2)

The 654th AC&W Squadron began operating a pair of AN/CPS-6B radars from this site in October 1951 and assumed coverage previously provided by a Lashup site at Grenier AFB, New Hampshire (L-4). An AN/FPS-8 was added in 1955. The Air Force eventually converted this unit to an AN/GPS-3 that served at Brunswick until the 1960s. In 1958 the AN/CPS-6Bs were retired and two AN/FPS-6 height-finder radars were installed. Brunswick was integrated into the SAGE system in 1959. This site was removed from service on March 1, 1965.

P-65/Z-65 — Charleston (A-3, A-9/GCI)

The 765th AC&W Squadron brought Charleston AFS to life in April 1952 and assumed coverage that had been provided by a Lashup site at Dow AFB (L-1). The site initially had AN/FPS-3 and 5 radars. In 1957 an AN/FPS-6 replaced the AN/FPS-5 height-finder radar. Another height-finder radar came in 1958 along with an AN/FPS-20 search radar that replaced the AN/FPS-3. During 1959 Charleston joined the SAGE system. In 1963 the site became the first in the nation to receive an AN/FPS-27. This radar subsequently was upgraded to become an AN/FPS-27A. The 765th was deactivated in September 1979.

LP-80/P-80/Z-80 — Caswell (A-3, A-5, A-9/SS)

During 1952 the 766th AC&W Squadron began operating a pair of AN/FPS-10 radars from this site and assumed coverage from a closing Lashup site at Limestone (L-50). In 1955 and 1956 an AN-FPS-8/GPS-3 was installed. The AN/GPS-3 remained in service until 1961. In 1957 and 1958 the AN/FPS-10s were phased out and two AN/FPS-6As arrived. Caswell came into the SAGE system in February 1959. In 1961 an ECCM-capable AN/FPS-7C began search duties. In 1979 Caswell became a TAC facility.

M-110/Z-110/J-54 — Bucks Harbor (A-6)

At the end of 1953 a station was sited at Bucks Harbor. This site originally was planned for Corea, Maine, but the Navy feared that the proposed site would interfere with a nearby radio interception and direction-finding facility. In 1956 the 907th AC&W Squadron began operating an AN/MPS-11 radar. In 1959 an AN/FPS-8 came to Bucks Harbor and this set subsequently was converted into an AN/GPS-3. In 1960 two AN/FPS-6A height-finder radars were activated and the site joined the SAGE system. By 1963 Bucks Harbor operated with an AN/FPS-24 search radar and AN/FPS-90 and AN/FPS-6B radars. Around 1965 Bucks Harbor became a joint-use facility. The 907th was deactivated in June 1979. Over the next two years preparations were made to turn over the facilities to the State of Maine and the FAA. Although the FAA assumed control of the operations area, the Air Force continued operating the AN/FPS-90 height-finder radar until 1988. In 1995 the search radar consisted of an AN/FPS-66A set.

Over-the-Horizon-Backscatter (OTH-B) — Moscow AFS and Columbia AFS

In the late 1970s GE Aerospace received a contract to build a prototype OTH-B radar system. A transmitter was built at Moscow AFS and a receiver was constructed at Columbia AFS. Testing of the prototype began in 1980. After two years of tests, GE Aerospace received a contract to expand the prototype into a full-scale model. Testing of the full-scale model began in 1988 and the Air Force accepted the site in 1990.

SAGE — Bangor

The Bangor Air Defense Sector was activated at Topsham AFS on January 8, 1957. The sector was disestablished in 1966. A SAGE direction center was operational here during the 1960s.

Maryland

L-14 — Fort Meade (A-1)

Starting in September 1950, this AN/CPS-6 Lashup radar guarded the approaches to Washington DC. Coverage was shifted to P-55 at Quantico, Virginia, in February 1952.

RP-54/Z-227 — Fort Meade (A-9/GCI)

This station became operational in 1962 with an AN/FPS-67 and two AN/FPS-6B radars. The 770th Radar Squadron moved here from former site P-54 located in Palermo, New Jersey. In 1963 one of the AN/FPS-6Bs was replaced by an AN/FPS-90. On July 1, 1963, the site was redesignated Z-227. Manned by the 770th Radar Squadron, this site came under TAC jurisdiction in 1979.

Massachusetts

L-5 — Otis AFB (A-1)

This Lashup site became operational in July 1950 using AN/CPS-5 and AN/TPS-10A radars. Coverage was assumed in October 1951 by site P-10 at North Truro.

P-10/Z-10/J-53 — North Truro (A-2, A-9/GCI)

The 762nd AC&W Squadron began operations with a pair of AN/CPS-3 radars at this Cape Cod site in 1951 and assumed radar coverage previously covered by a Lashup site at Otis AFB (L-5). In 1955 these units were joined by an AN/FPS-8 model. Eventually converted to an AN/GPS-3, this radar left service in 1960. The years 1958 and 1959 saw the arrival of AN/FPS-6 and 6A height-finder radars. During this time, North Truro was integrated into the SAGE system. In 1960 the 762nd started operating an AN/FPS-7 radar. In 1963 the height-finder radars were replaced by AN/FPS-26A and AN/FPS-90 sets. In 1979 the site came under TAC jurisdiction. In 1995 the FAA operated an AN/FPS-91A search set.

MM-1 — Fort Heath

This Army Missile Master (MM) site with an ARSR-1 radar supported Boston's anti-aircraft defense and was tied into the Air Defense Command and FAA networks in 1961. The site was removed from the ADC network in 1962.

PAVE PAWS — Otis AFB/Cape Cod AFS

Built to provide warning against submarine-launched ballistic missiles, this PAVE PAWS site became operational in 1979. The structure consists of two AN/FPS-115 phased-array radars mounted on a triangular building.

MICHIGAN

L-20 — Oscoda (A-1)

This Lashup site was operational from mid-1950 to mid-1951.

LP-16/P-16/Z-16/J-59 — Keweenaw/Calumet (A-2, A-5, A-9/NCC)

The 665th AC&W Squadron began operating AN/FPS-3 and AN/FPS-5 radars at this northern Michigan site in early 1953. In 1956 an AN/FPS-6 replaced the AN/FPS-5 height-finder radar. In 1958 an AN/FPS-20 search radar was deployed at this site. By 1961 this radar was upgraded and redesignated as an AN/FPS-64. A turnover of equipment in 1963 left the site with an AN/FPS-27 search radar along with AN/FPS-26A and AN/FPS-90 height-finder radars. Calumet came under TAC jurisdiction in October 1979.

L-17/LP-20/P-20/Z-20 — Selfridge AFB (A-1, A-2, A-5, A-9/SS)

Manned by the 661st AC&W Squadron, this Lashup site used an AN/CPS-5 radar. During a 1949 exercise this site performed well, tracking aircraft at ranges up to 210 miles. Selfridge became part of the permanent network in 1952. The site used a pair of AN/CPS-6 sets. An AN/FPS-6 height-finder radar arrived in 1957. Selfridge joined the SAGE system in 1959. By 1960 the AN/CPS-6 radars had been retired and were replaced by AN/FPS-20 search and AN/FPS-6 and 6A height-finder radars. In 1961 the site received an AN/FPS-35 radar; however, problems prevented the radar from becoming operational in 1961. It became operational the following year. Meanwhile an AN/FPS-26A replaced the AN/FPS-6A height-finder radar. The 661st was deactivated in July 1974.

P-34/Z-34/J-58 — Empire (A-2, A-9/GCI)

In late 1951 the 752nd AC&W Squadron began operating a pair of AN/CPS-6B radars at this site. In 1958 one of these radars was replaced by an AN/FPS-6. A second height-finder radar arrived a year later. In 1960 this site began operating an AN/FPS-7 radar and joined the SAGE system. One of the AN/FPS-6 height-finder radars was supplanted by an AN/FPS-26A in 1963. In 1964 this site became an ADC/FAA joint-use site. The 752nd was deactivated in April 1978. FAA assumed control of the operations area except for the height-finder radar. The National Park Service assumed control of the rest of the property.

LP-61/P-61/Z-61/J-57 — Port Austin (A-3, A-5, A-9/NCC)

The 754th AC&W Squadron began operations at this site in June 1952. The site used an AN/FPS-3 set. An AN/CPS-4 was added in 1954. This radar was replaced in 1957 by an AN/FPS-6 height-finder radar. A second height-finder radar was added a year later. In 1958 this site began operating an AN/FPS-20 radar. SAGE operations began in August 1959. In early 1962 the site received and operated an AN/FPS-24. This site came under TAC control in October 1979.

L-21/LP-66/P-66/Z-66 — Sault Ste. Marie (A-1, A-2, A-5)

An AN/TPS-1C Lashup radar was operational at this site in 1950. This radar was later incorporated into the permanent network. In 1952 the 753rd AC&W Squadron operated AN/FPS-3 and AN/FPS-5 units. During 1956 the Air Force replaced the AN/FPS-5 with an AN/FPS-6 unit. In 1958 this site was operating an AN/FPS-20 radar and soon received a second height-finder radar. In 1960 the site provided data to the regional SAGE command center. In 1963 an AN/FPS-35 replaced the AN/FPS-20A. This facility came under TAC jurisdiction in October 1979.

LP-67/P-67/Z-67 — Fort Custer/Custer (A-3, A-5)

The 781st AC&W Squadron was operating AN/FPS-3 and AN/CPS-4 radars at this site as of April 1952. An AN/FPS-4 replaced the AN/CPS-4 in 1956 and an AN/FPS-6 superseded this unit two years later. Also in 1958 an AN/FPS-20 replaced the AN/FPS-3 search radar. This radar was upgraded to an AN/FPS-66 in 1961. A second height-finder radar was installed in 1959. Also in 1959, Custer joined the SAGE system. The Air Force removed Custer from service on March 1, 1965.

M-105 — Alpena (A-6)

Beneficial occupancy was achieved at this Phase I mobile radar site in late 1954. Operations began in 1956. Budget cuts forced the station to close in 1957.

M-109 — Grand Marais (A-6)

Beneficial occupancy was achieved at this Phase I mobile radar site in late 1954. Operations began in 1956. Budget cuts closed the station in 1957.

SAGE — Fort Custer

The Detroit Air Defense Sector was activated on January 8, 1957. A SAGE direction center operated here during the 1960s.

SAGE — Sault Ste. Marie

The Sault Ste. Marie Air Defense Sector was activated on November 8, 1958. A SAGE direction center was active at K. I. Sawyer Airport during the 1960s.

Minnesota

P-17/Z-17 — Leaf River/Wadena (A-4, A-9/SS)

The 739th began operating AN/FPS-3 and AN/FPS-4 sets at this site in June 1952. The AN/FPS-4 unit was superseded in 1956 by a AN/FPS-6 height-finder radar. In 1958 this site was operating an AN/FPS-20 radar. 1959 saw the integration of Wadena into the SAGE system and the arrival of a second height-finder radar (AN/FPS-6A). In 1961 this set was upgraded and redesignated as an AN/FPS-64. During 1963 one of the height-finder radars was supplanted by an AN/FPS-90. The Air Force deactivated the 739th in September 1970.

P-18/Z-18 — Moulton/Chandler (A-4, A-9/SS)

In June 1952 the 787th AC&W Squadron began operations with AN/FPS-3 and AN/FPS-4 sets. By 1959 these sets had been replaced by AN/FPS-20 and AN/FPS-6 radars. In 1961 the AN/FPS-20 set was upgraded and redesignated as an AN/FPS-64 and a second height-finder radar (AN/FPS-26) was added. By 1966 this site was operating an AN/FPS-27 set. The Air Force deactivated the 787th in September 1969.

LP-69/P-69/Z-69 — Finland (A-2, A-5, A-9/NCC)

In July 1952 the 756th AC&W Squadron was operating AN/FPS-3 and AN/FPS-5 radars from this site. By 1959 these radars had been replaced with AN/FPS-20 and AN/FPS-6 sets and a second height-finder radar (AN/FPS-6A) was being installed. In November 1959 Finland joined the SAGE system. In 1961 the search radar was upgraded and redesignated as an AN/FPS-64. In 1963 the height-finder radars were replaced by AN/FPS-26A and AN/FPS-90 sets. By 1966 there was an AN/FPS-27 operating at this site. The site came under TAC jurisdiction in 1979.

M-101 — Rochester (A-6)

The 808th AC&W Squadron began operations in 1956 with a AN/TPS-1D radar. Due to budget cuts, it was removed from service in 1957.

SM-132/Z-132 — Baudette (A-7, A-9/GCI)

In 1959 the 692nd AC&W Squadron began operating an AN/FPS-3 and a pair of AN/FPS-6 height-finder radars at this site. In 1962 this SAGE center received an

AN/FPS-24 radar. During the following year, AN/FPS-26 and AN/FPS-90 radars assumed height-finder duties. In 1979 the site came under TAC jurisdiction.

SM-138 — Grand Rapids (A-7)

Grand Rapids became operational in 1957 with an AN/FPS-3 radar. In 1958 the 707th AC&W Squadron was operating AN/FPS-20 and AN/FPS-6 sets. A year later Grand Rapids joined the SAGE system. In 1961 the AN/FPS-20A was upgraded and redesignated as an AN/FPS-67. In March 1963 the Air Force ordered the site to close. Operations ceased on May 1.

SM-139 — Willmar (A-7)

This site became operational in 1957 and closed in 1960 due to budgetary constraints. The 721st AC&W Squadron operated a variety of radars at Willmar including AN/FPS-6, AN/FPS-8, AN/MPS-7, and AN/FPS-3 sets.

Z-306/J-60 — Nashwauk

Constructed in 1980, Nashwauk was a JSS location where the FAA served as host and the Air Force was the tenant operating a height-finder radar. Air Force operations ceased in 1987.

SAGE — Duluth

A SAGE direction center at Duluth Municipal Airport served the Duluth Air Defense Sector during the 1960s.

Mississippi

TM-195/Z-195 — Crystal Springs (A-8)

This site was placed in operation by the 627th AC&W Squadron in late 1959 as a test site for the AN/FPS-27 radar. This radar became operational in the 1960s. The 627th was deactivated in September 1968.

Missouri

P-64/Z-64 — Sublette/Kirksville (A-4, A-9/SS)

The 790th AC&W Squadron activated a pair of AN/FPS-10 radars at this site in April 1952. One AN/FPS-10 height-finder radar was phased out in 1958 with the arrival of two AN/FPS-6A sets. The other AN/FPS-10 stayed in operation until 1963 and then was replaced with an AN/FPS-7. Kirksville began SAGE operations in 1959. The Air Force deactivated the 790th in September 1968.

P-68 — Fordland (A-4)

The 797th AC&W Squadron began operating AN/FPS-3 and AN/FPS-4 radars from this site in 1952. During 1954 an AN/TPS-10D saw service here. In 1958 an AN/FPS-6 replaced the AN/FPS-4 height-finder radar. This site was deactivated in 1960 due to budget constraints.

SAGE — Kansas City

A SAGE Combat Operations Center and Air Defense Regional Headquarters was located at Richards-Gebaur AFB during the 1960s.

Montana

P-11 — Bonner's Ferry/Yaak (A-4)

The 680th AC&W Squadron began operating AN/FPS-3 and AN/FPS-4 sets at this northern tier site in April 1952. An AN/FPS-6 was added in 1956 and the AN/FPS-4 was replaced by an AN/GPS-3 in 1957. This station at Yaak was converted into an unmanned gap-filler radar site in 1960 and redesignated SM-151E. The 680th AC&W Squadron was deactivated on July 1, 1960 but would be reborn the following year at site P-54 in New Jersey.

P-24/Z-24 — Del Bonito/Cut Bank (A-4)

The 779th AC&W Squadron started operating AN/FPS-3 and AN/FPS-4 radars in April 1952. In 1958 an AN/FPS-20 search radar replaced the AN/FPS-3 at this site. In the following year two AN/FPS-6A height-finder radars superseded the AN/FPS-4. In 1961 this site was integrated into the SAGE system. In 1962 the AN/FPS-20A was further upgraded and redesignated as an AN/FPS-66. This site was removed from service on March 1, 1965.

Havre Air Force Station, Montana: Life

The Air Force considered Havre AFS a remote duty site. The airmen at the station must have agreed with this for they named the station newspaper the Isolation Times. Despite its isolated location, retired personnel remembered Havre as a pleasant duty station. The Air Force worked hard to keep its personnel entertained. The station personnel services department, for example, described the surrounding countryside as a "hunter's and fisherman's paradise," and kept a full range of hunting and fishing gear, softball, and tennis equipment on hand. It even had boats and a jeep available for weekend excursions. In addition, the station also had a very active non-commissioned officers club; a two-lane bowling alley; hobby shops for woodworking; photography; and auto repair; a library; movie theater; barber shop; and exchange and commissary. If the men wanted to go into town, Havre was an hour drive down Route 232.

The addition of family housing, built for both officers and senior enlisted men, also brought a degree of stability and permanence to the station. Grade school children attended a new two-room school built just outside the station. The Air Force bussed the older students into Havre to attend school there.

P-25 — Simpson/Havre (A-4, A-9/NCC)

In 1952 the 778th AC&W Squadron began operating AN/FPS-3 and AN/FPS-4 radars at this site. The Air Force added an AN/GPS-3 in 1956 that stayed until 1962. The site was integrated into the SAGE system in 1961. The AN/FPS-3 remained operational into the early 1960s while two AN/FPS-6 models took over height-finder duties. The 778th was deactivated in September 1979.

P-26/Z-26 — Opheim (A-4, A-9/SS)

The 779th AC&W Squadron began operations in 1952 with AN/FPS-3 and AN/FPS-4 radars. An AN/GPS-3 radar operated here between 1957 and 1961. In 1958 and 1959 AN/FPS-6 and AN/FPS-6A radars replaced the AN/FPS-4 height-finder radar. The

---

HAVRE AIR FORCE STATION, MONTANA: COMMUNITY RELATIONS

Interaction between the radar station and local residents played out on two different levels with the city of Havre and with the farmers and ranchers who lived near the station.

To the people of Havre, the "radar base," as they called it, was a welcome addition to the community and did not appreciably diminish the character of the town. Many Air Force families lived in Havre and their children attended school there. They took an active role in scouting, coached soccer, taught karate, and belonged to the Lions and Rotary clubs. People from the station often shopped in Havre and on weekends frequented the town's restaurants and bars. The station bought some of its supplies from town merchants and employed a dozen local residents as plumbers, carpenters, and boiler technicians.

The town of Havre worked diligently to foster good relations with the Air Force. The Havre Area Chamber of Commerce sponsored a very active Military Affairs Committee that sought to bring the town and the station together. The Chamber of Commerce and area merchants sponsored a military discount program and also assisted military families in finding housing. The Air Force did its part too. The station held frequent open houses, fielded a team in the Havre softball league, and participated in various civic projects.

The radar station forged even closer ties with the farmers and ranchers who lived on the sparse prairie around the base. For them the station was the largest community, and the only bar, within twenty miles. The local farmers and ranchers were always welcomed at the station's NCO club and also participated in many other social and athletic events at the station. The station also stood ready to assist its neighbors during emergencies. The farmers were also willing to lend a hand. On more than one occasion they pulled government vehicles out of muddy ditches or rescued them from flooded roads.

AN/FPS-3 left service in 1960 and was replaced by an AN/FPS-7C FD radar. In 1961 the site was integrated into the SAGE system. In 1963 the 779th Radar Squadron operated the AN/FPS-7C search radar along with AN/FPS-26A and AN/FPS-90 height-finder radars. The 779th was deactivated in September 1979.

M-98/Z-98 — Miles City (A-6, A-9/NCC)

Beneficial occupancy was achieved at this Phase I Mobile Radar site in late 1954. By August 1955 the 902nd AC&W Squadron was operating AN/MPS-7 and AN/MPS-14 radars. In 1958 an AN/FPS-20 radar replaced the AN/MPS-7. In 1961 this set was upgraded and redesignated as an AN/FPS-66. Also in 1961 an AN/FPS-6 height-finder radar was installed and the site joined the SAGE system. Later in the decade, the site received an AN/FPS-27. The 902nd was deactivated in June 1968.

SM-147/Z-147/J-77 — Malmstrom AFB (A-7)

Under the control of the 801st AC&W Squadron, Malmstrom became operational with AN/FPS-20 and AN/FPS-6 radars in 1957. A second height-finder radar was added in 1960 and subsequently was upgraded to an AN/FPS-90 set. In 1959 this station was performing air traffic control duties for the FAA. Two years later, Malmstrom was a SAGE center. By 1966 the site hosted an AN/FPS-24 radar. The 801st was deactivated in December 1969.

TM-178/Z-178 — Lewiston/Lewistown (A-8, A-9/SS)

The 694th AC&W Squadron began operations in 1960 with AN/FPS-3A and AN/FPS-6 radars. A second height-finder radar was added a year later. Also in 1961 Lewiston began feeding the SAGE system. By 1966 the 694th was operating an AN/FPS-66A search set. The Air Force deactivated the 694th in June 1971.

TM-179/Z-179/J-78 — Kalispell (A-8, A-9/SS)

The 716th AC&W Squadron started operations with an AN/FPS-7 and a pair of AN/FPS-90 radars in 1961. The 716th was deactivated in April 1978.

SAGE — Malstrom AFB, Great Falls

A Great Falls Air Defense Sector was activated on March 1, 1959 at Malmstrom AFB. There was a SAGE direction center located here during the 1960s.

Nebraska

P-71/Z-71 — Omaha (A-4, A-9/SS)

The 789th AC&W activated an AN/CPS-4 and AN/FPS-3 at this site in April 1952. Eventually the Air Force replaced the height-finder radar with an AN/FPS-6 and replaced the AN/FPS-3 search radar with an AN/FPS-20. In late 1959 this station was also performing air traffic control duties. SAGE operations began in 1961. A second height-finder radar (AN/FPS-6A) was installed in 1962. In the early 1960s the AN/FPS-20A radar was upgraded to become an AN/FPS-66A. The 789th was deactivated in September 1968.

SM-133/Z-133 — Hastings (A-7, A-9/SS)

Deployment of an AN/FPS-67 search radar and two AN/FPS-6 height-finder radars to Hastings in 1962 marked the completion of the second phase of the mobile radar program. This SAGE feeder station was manned by the 625th Radar Squadron. This squadron was deactivated in September 1968.

Z-217 — North Platte

This ARSR-2 FAA joint-use radar began feeding information into the air defense network in 1963.

NEVADA

M-127/Z-127 — Winnemucca (A-6, A-9/NCC)

This site typified some of the problems with the mobile radar program. Because the program received minimal funding, the cantonment area was sited on Department of the Interior land located several miles away from the radar, rather than on adjacent land owned by the Southern Pacific Land Company. Delays in the decision-making process set the date for beneficial occupancy back to February 1955. Operational status was finally achieved by the 658th AC&W Squadron in 1956. The site used an AN/FPS-3 radar. By 1959 this radar had been joined by a pair of AN/FPS-6B height-finder radars. In 1960 the AN/FPS-3 was replaced by an AN/FPS-20 search set. At the end of 1961 this search set had been upgraded into an AN/FPS-66. At this time Winnemucca was providing a feed into the SAGE system. In 1963 the two height-finder radars were converted to AN/FPS-90 units. The 658th was deactivated in June 1968.

SM-156/Z-156 — Fallon NAS (A-7, A-9/SS)

This site became operational in 1956 when the 858th AC&W Squadron activated AN/MPS-7 and AN/MPS-14 sets. An AN/FPS-3 search set briefly saw service here in 1959. Around 1965 it became an FAA/ADC joint-use facility. The site used an AN/FPS-35 search radar that replaced the AN/MPS-7 set in 1963. The 858th was deactivated in June 1975.

SM-163/Z-163 — Las Vegas (A-7, A-9/SS)

In 1956 the 865th AC&W Squadron activated AN/FPS-3 and AN/MPS-14 radars at this site. In 1958 an AN/FPS-20A replaced the AN/FPS-3 search radar. By 1961 Las Vegas was an ADC/FAA joint-use facility and provided data for the SAGE system. An AN/FPS-26 height-finder radar joined the site in 1963. The search radar was replaced in the mid-1960s with an AN/FPS-27 set. The 865th was deactivated in December 1969.

SM-164/Z-164 — Tonopah (A-7)

This site became operational in 1957 when the 866th AC&W Squadron activated an AN/MPS-7 radar. In June 1961 the site was moved to another peak. At the new location, the 866th operated a pair of AN/FPS-6 height-finder and AN/FPS-7C search radars as part of the SAGE system. In 1963 the height-finder radars were converted to AN/FPS-90 sets. The Air Force deactivated the 866th in September 1970.

Z-214 — Battle Mountain

This FAA ARSR-2 radar began feeding information into the air defense network in 1963.

SAGE — Reno

The Reno Air Defense Sector was activated on February 15, 1959, and disestablished in 1966. A SAGE direction center controlled this sector from Stead AFB.

New Hampshire

L-4 — Grenier AFB (A-1)

This Lashup site became operational with an AN/CPS-5 radar in June 1950. In February 1951 coverage was assumed by site P-13 at Brunswick, Maine.

M-104 — Rye (A-6)

In 1956 the 644th AC&W Squadron activated an AN/TPS-1. Budget cuts forced the Air Defense Command to close this station in 1957.

NEW JERSEY

L-12/LP-9/P-9/Z-9 — Twin Lights/Navesink/Highlands (A-1, A-2, A-5, A-9/SS)

In 1948 the 646th AC&W Squadron activated a pair of AN/CPS-6 radars at this coastal site to feed into a primitive control center established at Roslyn, New York. During an exercise in mid-1949, the capability of this site was judged as "zero." These radars were incorporated into the Lashup system and the follow-on permanent network. In 1955 the site received an AN/FPS-8 radar. This radar was converted into a AN/GPS-3 that would remain until 1960. In 1958 an AN/FPS-6 height-finder radar became operational. Also that year the Navesink complex began providing a feed into the SAGE blockhouse located at McGuire AFB. In September 1959 this site became the first to deploy an AN/FPS-7 radar. In 1960 the Air Force installed an AN/FPS-6B height-finder radar. By 1963 the AN/FPS-6 and 6B height-finder radars had been replaced by AN/FPS-26A and AN/FPS-90 sets. The site was removed from service on April 1, 1966.

L-13/LP-54/P-54/Z-54 — Palermo (A-1, A-3, A-5, A-9/NCC)

In 1948 the Air Force activated an AN/TPS-1B long-range search radar at this southern New Jersey site that fed into a primitive control center established at Roslyn, New York. This radar site was incorporated into the Lashup system and the follow-on permanent network. In 1951 AN/CPS-5 and AN/TPS-10A height-finder radars were added to the site. By April 1952 the 770th AC&W Squadron was operating AN/CPS-4 and AN/FPS-3 sets. In the spring of 1957, this site was one of the first to deploy an AN/FPS-20 radar. The site also received two AN/FPS-6 height-finder radars at this time. The site became a SAGE feeder in June 1958. By late 1959 this station also was performing air traffic control duties. On October 1, 1961, the 770th and the site designation left for Fort Meade, Maryland (RP-54). The Palermo site was then operated by Detachment 1 of the New York Air Defense Sector. This unit was redesignated the 680th Radar Squadron in 1962 and the P-54 site designation returned to Palermo in 1963. In 1963, the AN/FPS-20 was upgraded into an AN/FPS-65. The 680th was deactivated in May 1970.

RP-63/Z-63/J-51 — Gibbsboro

Initially this site was operating during the late 1950s as a gap-filler radar site with an AN/FPS-14 radar. To provide more protection for the east coast, the Air Force moved site designation P-63 and the 772nd Radar Squadron from Claysburg, Pennsylvania to this site located twelve miles east of Philadelphia. This station began operations in 1961 with AN/FPS-66 and AN/FPS-6 radars. The center initially fed data to the SAGE center at McGuire AFB. An AN/FPS-27 replaced the AN/FPS-66 in 1963. At this time the site used AN/FPS-6A and AN/FPS-26A height-finder radars. The site became an ADC/FAA joint-use facility around 1966. The 772nd continued operating the AN/FPS-27 and AN/FPS-26A radars until 1983. The squadron was deactivated in 1984. However, an AN/FPS-117 became operational in 1986 under contractor support. This radar served as a prototype for units installed in Alaska. After shutdown in 1992 this radar was moved to Murphy's Dome, Alaska. A portion of the site was transferred to the FAA in 1995 for installation of an ARSR-4 set.

SAGE — McGuire AFB

A SAGE direction center located here served the New York Air Defense Sector during the 1960s.

NEW MEXICO

LP-7/P-7 — Gonzales/Continental Divide (A-2, A-5)

The 769th AC&W Squadron began operations in 1952. In 1953 the radars consisted of AN/FPS-3 and AN/FPS-5 sets. In 1955 an AN/FPS-4 height-finder radar was installed, only to be replaced a year later by an AN/FPS-6 model. At the end of 1960, this radar was transferred to the FAA pending deactivation of the 769th on June 28, 1961.

L-44/LP-8/P-8 — Los Alamos/El Vado/Tierra Amarillo (A-1, A-2, A-5)

An AN/CPS-5 Lashup radar was activated in 1950. The unit was later installed at the permanent site that was activated to provide coverage for Los Alamos. By September 1952 the 767th AC&W Squadron commenced operation at this new permanent site. By 1953 AN/FPS-3 and AN/FPS-5 radars were in operation. A year before the site shutdown the AN/FPS-5 was replaced by an AN/FPS-6 height-finder radar. On December 8, 1958, this became the first radar of the permanent network to shutdown as the site was poorly located.

LP-51/P-51 — Moriarity (A-2, A-5)

By September 1952 the 768th AC&W Squadron operated AN/FPS-3 and AN/FPS-5 radars from this site. Six years later an AN/FPS-20 radar replaced the AN/FPS-3 search radar and an AN/FPS-6 performed height-finding duties. In 1960 this site was transferred to the FAA pending deactivation of the 768th.

L-46/M-90 — Walker AFB (A-1, A-6, A-9/SS)

An AN/CPS-5 Lashup radar was in place at this site in 1950 to protect the approaches of Walker AFB. Designated to receive a new radar as part of the mobile radar program, this radar site continued to be operational on a Lashup basis in late 1952 using an AN/TPS-1B radar. A more permanent facility at Walker was operational by late 1955. At this time the 686th AC&W Squadron operated AN/MPS-7 and AN/MPS-14 radars. In March 1963 the Air Force ordered the site to shut down. Operations ceased two months later.

L-45/M-94/Z-94 — Kirtland AFB/West Mesa (A-1, A-6, A-9/GCI)

A Lashup station was operational here by late 1949. The station provided coverage for the Albuquerque-Los Alamos region. The site used AN/CPS-5 and AN/CPS-4 radars. In late 1952 an AN/CPS-5 radar was still operating on a Lashup basis to provide temporary coverage. Meanwhile plans to permanently site the radar on Mt. Vulcan were thwarted by an owner who refused to sell at the price offered. Instead the Air Force selected a site twenty-two miles west of Albuquerque. The site became operational in 1956. At this site the 687th AC&W Squadron used AN/MPS-7 and AN/MPS-14 radars. In 1960 this site was also handling air traffic control functions for the FAA. In 1961 the site hosted an AN/FPS-20 search radar. By 1966 the 687th was operating an AN/FPS-91A search set. The Air Force deactivated this squadron in September 1968.

M-95 — Las Cruces (A-6, A-9/NCC)

In 1955 the 685th AC&W Squadron began operating an AN/MPS-7 radar at this site. Over the next two years equipment additions and deletions included AN/MPS-8, AN/TPS-1D, and AN/MPS-14 sets. By 1958 the 685th was operating AN/FPS-20 and AN/MPS-14 radars. In March 1963 the Air Force ordered the site to shut down; operations ceased in April.

Z-221 — Gallup

Operated by the FAA, this ARSR-2 radar began providing information to the air defense network in 1963.

Z-245/J-28 — Silver City/Cliff

In late 1972, a detachment of the Luke-based 4629th Air Defense Squadron arrived to set up and operate an AN/FPS-6 height-finder radar in conjunction with the FAA ARSR-2 set. The site continued as an FAA-operated JSS site after the 4929th was deactivated in 1975.

NEW YORK

L-10/LP-45/Z-45 — Camp Hero/Montauk (A-1, A-3, A-5, A-9/GCI)

An AN/TPS-1B long-range search radar was activated at this site in June 1948. This site fed into a primitive control center established at Roslyn. This site was incorporated into the Lashup and subsequent permanent network with the 773rd AC&W Squadron overseeing the facility. In 1951 AN/CPS-5 and AN/TPS-10A height-finder radars were placed on the site. A year later AN/FPS-3 and AN/FPS-5 radars were operating. Between 1955 and 1956 an AN/FPS-8/GPS-3 made an appearance at the tip of the Long Island site. In the spring of 1957 this site received one of the first AN/FPS-20 units along with a pair of AN/FPS-6 height-finder radars. During 1958 Montauk began SAGE operations. In December 1960 the first of the specific frequency diversity radars, an AN/FPS-35, became operational at Montauk. This powerful radar caused radio interference problems in the vicinity. These problems caused this radar to be taken out of service in 1961. With the problems resolved, the radar was operational again in 1962 and by 1963 an AN/FPS-26 had replaced one of the AN/FPS-6 height-finder radars. In 1963 the site also had become an FAA/ADC joint-use facility. Around 1965 the site was removed from joint-use status. Montauk came under TAC jurisdiction in 1979. The facility was decommissioned in the early 1980s. A site at Riverhead (Z-315/J-52) assumed coverage.

L-6 — Pine Camp (A-1)

Pine Camp became operational using an AN/CPS-5 radar in June 1950. In 1951 an AN/TPS-10A radar became operational. In June 1952 coverage was assumed by site P-49 at Watertown.

L-7 — Schenectady (A-1)

This Lashup site was activated in 1950. The site used both the AN/CPS-5 and AN/TPS-10A radars. Operations continued until February 1952 when coverage was assumed by site P-50 at Schuylerville.

L-8 — Seneca (A-1)

This Lashup site at the Ordnance Depot used AN/CPS-5 and AN/TPS-10A radars. Operations lasted from June 1950 until coverage was assumed by site P-14 at Bellevue Hill, Vermont, in September 1951.

L-11 — Santini (A-1)

This Lashup site was activated in May 1950. The site used an AN/CPS-6 medium-range radar. It was deactivated in October 1951.

L-19 — Fort Niagara (A-1)

This Lashup site was activated with an AN/CPS-6 radar in August 1950. Coverage was assumed in September 1951 by site P-21 at Shawnee.

P-21/Z-21 — Shawnee, Lockport (A-2, A-9/SS)

The 763rd AC&W Squadron began operating a pair of AN/CPS-6Bs from this site in September 1951, taking over for the Lashup radar at Fort Niagara (L-19). These radars were phased out at the end of the decade. Meanwhile SAGE operations began at this site in December 1958. This site started operating an AN/FPS-7 radar in 1960. Height-finding chores were handled at that time by a pair of AN/FPS-6s. During 1962 an AN/FPS-26 replaced one of these height-finder radars. The 763rd was deactivated in June 1979.

P-49/Z-49 — Watertown (A-2, A-9/NCC)

In June 1952 the 655th AC&W Squadron started operating AN/FPS-3 and AN-FPS-5 radars and assumed coverage previously provided by site L-6 at Pine Camp. In 1958 this site was operating with AN/FPS-20 and AN/FPS-6 sets and had joined the SAGE system. A second AN/FPS-6 height-finder radar was added during 1959. In 1961 the search radar was upgraded and redesignated as an AN/FPS-66. A height-finder radar was converted to an AN/FPS-26A in 1963. By 1966 the site operated an AN/FPS-27. The site came under TAC jurisdiction in 1979.

P-50/Z-50 — Schuylerville/Sarasota Springs (A-3, A-9/NCC)

The 656th AC&W Squadron began operating AN/FPS-3 and AN/FPS-5 radars from this central New York site in April 1952. This site assumed coverage from site L-7 in Schenectady. This site was one of the first to receive an AN/FPS-20 search radar in the spring of 1957. By 1958 the original radars were gone and operations continued with an AN/FPS-20 and a pair of AN/FPS-6 height-finder radars. At this time, Sarasota Springs began providing information to the SAGE system. The search radar subsequently was upgraded in 1961 to become an AN/FPS-65. Major equipment changes occurred during 1963 with the arrival of an AN/FPS-27 search radar along with AN/FPS-26A and AN/FPS-90 height-finder radars. In 1964 this site became an ADC/FAA joint-use facility. The 656th was deactivated in June 1977.

Z-315/J-52 — Riverhead/Suffolk

This JSS site was constructed in the early 1980s to replace Montauk (Z-45). The Air Force operated the height-finder radar on this FAA site.

Z-312/J-55 — Star Hill/Utica

In 1974 the FAA constructed a radar site on Star Hill. The Air Force joined the FAA at this location in 1979 when a height-finder tower was constructed. With this feature, Utica became part of the JSS network. In the late 1980s the Air Force ceased operating the height-finder radar.

J-56 — Dansville

This FAA radar provided a feed into the JSS starting in the 1980s.

SAGE — Hancock Field

This SAGE site served as a direction center for the Syracuse Air Defense Sector in the 1960s. The Regional Headquarters and the Region Combat Center was collocated here.

SAGE — Stewart AFB

This SAGE direction center at Stewart AFB controlled the Boston Air Defense Sector during the 1960s.

North Carolina

M-115/Z-115/J-02 — Fort Fisher, Cape Fear (A-6, A-9/GCI)

This site on Cape Fear was operational in 1955 when the 701st AC&W Squadron activated AN/MPS-7 and AN/MPS-8 radars. By 1958 the AN/MPS-8 had been converted into an AN/GPS-3 and an AN/MPS-14 had been added. In 1962 an AN/FPS-7C and AN/FPS-26 were placed in operation along with the AN/MPS-14. Fort Fisher began feeding into the SAGE system in 1963. The site came under TAC jurisdiction in 1979. Upon decommissioning the site, the FAA continued to run the radar as part of the JSS network. In 1995 an AN/FPS-91A performed search duties.

M-116 — Cherry Point Marine Corps Air Station (MCAS) (A-6, A-9/SS)

In 1958 the 614th AC&W Squadron activated an AN/FPS-6 and a pair of AN/FPS-8 radars at this Marine Corps Air Station. Cherry Point ceased operations on April 30, 1963.

M-117/Z-117 — Roanoke Rapids (A-6)

This site became operational in 1956 when the 632nd AC&W Squadron activated AN/MPS-11 and AN/TPS-10D sets. Two years later, the AN/TPS-10D was replaced with a pair of AN/FPS-6 height-finder radars. In 1959 Roanoke Rapids joined the SAGE system. In 1966 an AN/FPS-27 was performing search duties at this site. The 632nd was deactivated on September 30, 1978.

M-130/Z-130 — Winston-Salem (A-6, A-9/GCI)

Funding shortfalls in FY 1957 forced the Air Defense Command to hold up construction at this site. However, the 810th AC&W Squadron did start operating AN/MPS-11 and AN/FPS-6A radars in late 1957. The AN/MPS-11 set was replaced by an AN/FPS-8 set that saw service between 1960 and 1962. In 1962 the 810th began operating AN/FPS-24 search and AN/FPS-26 height-finder radars along with the previously installed AN/FPS-6 set. In 1963 Winston-Salem began providing data for the SAGE system. The Air Force deactivated the 810th in July 1970.

North Dakota

P-27/Z-27 — Fortuna (A-4, A-9/GCI)

In April 1952 the 752nd AC&W Squadron began operations at this site using AN/FPS-3 and AN/FPS-4 radars. During 1957 an AN/GPS-3 made a brief appearance. In 1958 the 752nd began operating an AN/FPS-20 search radar that replaced the AN/FPS-3 and the AN/FPS-6 that had superseded the AN/FPS-4 height-finder radar. By 1960 a pair of AN/FPS-6As handled height-finder chores. During 1961 Fortuna became part of the SAGE system. By 1963 the makeup of the radars at Fortuna consisted of an AN/FPS-35 search radar along with AN/FPS-6 and AN/FPS-90 height-finder radars. The 780th was deactivated in September 1979.

P-28/Z-28 — Velva/Minot (A-4)

The 786th AC&W Squadron began operations using AN/FPS-3 and AN/FPS-5 radars in April 1952. During 1957 an AN/GPS-3 saw brief use. In 1958 the original radars were replaced by AN/FPS-20 search and AN/FPS-6 height-finder sets. A second height-finder set (AN/FPS-6B) was installed during the following year. SAGE operations began in 1961. By the end of 1961 the search set had been upgraded and redesignated as an AN/FPS-66. By 1966 the site was operating an AN/FPS-27. The 786th was deactivated in September 1979.

P-29/Z-29/Z-303/J-75 — Finley (A-4, A-9/NCC)

The 785th AC&W Squadron began operating an AN/FPS-3 and AN/FPS-5 from this site in April 1952. In 1957 an AN/FPS-6 replaced the AN/FPS-4 height-finder radar and an AN/FPS-8 saw brief service. At the end of 1958 this site was operating an AN/FPS-20 radar. A second height-finder was added during 1959. At the end of 1959, Finley became a SAGE system input provider. In 1962 this station went operational with an AN/FPS-35 search radar. During 1963 this set was complemented by AN/FPS-6 and AN-FPS-90 height-finder radars. The site was transferred to the TAC in October 1979.

TM-177/Z-177 — Dickinson (A-8, A-9/SS)

This site became operational in 1959 with AN/FPS-3 and AN/FPS-6A radars manned by the 706th AC&W Squadron. A second AN/FPS-6A height-finder radar was added in 1960. By 1963 these radars had been upgraded to AN/FPS-66 and AN/FPS-90 units. Dickinson was removed from service on March 1, 1965.

Searching the Skies: The Legacy of the United States Cold War Defense Radar Program

Z-300/J-76 — Watford City

The Air Force built a height-finder radar tower at this FAA site in 1979. Height-finder operations at this FAA-owned JSS site ceased in the mid-1980s. As of 1995 the FAA operated an AN/FPS-67B search set.

SAGE — Grand Forks AFB

A SAGE direction center operated at Grand Forks AFB during the 1960s.

SAGE — Minot AFB

A SAGE direction center operated at Minot AFB during the 1960s.

PARCS — Concrete

The Perimeter Acquisition Radar attack Characterization system was constructed here from 1969 through 1974 as a component of the ABM system. With the cancellation of that program, the large radar was transferred to the Air Force for space-tracking duties.

Ohio

L-18 — Ravanna (A-1)

This Lashup site located at the airport used AN/CPS-5 and AN/TPS-10A radars. Operations lasted from May 1950 until April 1952 when site P-62 at Brookfield assumed area coverage.

L-22 — Lockbourne AFB (A-1)

This Lashup site was operational in 1950, using an AN/TPS-1B radar. In April 1952 site LP-73 at Bellefontaine assumed coverage of the area.

P-62 — Brookfield (A-3)

The 662nd AC&W Squadron began operating AN/FPS-3 and AN/FPS-5 radars from this site in April 1952. This operation allowed site L-18 at Ravanna to shut down. Eventually the height-finder radar was replaced with an AN/FPS-4 and then with an AN/FPS-6. This site at Brookfield became a gap-filler radar site (RP-62E). The P-62 site designation was transferred to Oakdale, Pennsylvania.

P-73/Z-73 — Bellefontaine (A-3, A-9/GCI)

A replacement site for L-22 at Lockbourne, this site became active in 1952 when the 664th AC&W Squadron turned on AN/FPS-3 and AN/CPS-4 radar sets. In 1958 the 664th was operating AN/FPS-20 search and AN/FPS-6 and 6A height-finder radars. SAGE operations began in August 1959. The search radar was upgraded and redesignated as an AN/FPS-66 in 1961. By 1966 the site was using an AN/FPS-27. The 664th was deactivated in September 1969.

Oklahoma

P-52/Z-52 — Tinker AFB/Tuttle/Oklahoma City (A-4, A-9/NCC)

The 746th AC&W Squadron began operating a pair of AN/FPS-10 radars from this site in May 1952. An AN/FPS-6 height-finder radar was added in 1958. The AN/FPS-10s were phased out, with the last one being removed in 1962. In 1963 this was an ADC/FAA joint-use facility. The site used AN/FPS-67 and AN/FPS-6 radars. The Air Force deactivated the 746th in September 1968.

P-77 — Bartlesville (A-4)

The 796th AC&W Squadron began operating a pair of AN/FPS-10 radars at this site in May 1952. In the mid-1950s an AN/FPS-6 replaced the AN/FPS-10 height-finder radar. This site was deactivated in 1960 due to budgetary constraints.

SAGE — Oklahoma City

A Region Headquarters and Region Combat Center was located here during the 1960s.

OREGON

L-33 — Portland AFB

An AN/TPS-1B radar was installed and activated on a 24-hour-a-day basis in early 1948 to protect the air approaches to Hanford, Washington. At that time the site was incorporated into the Lashup system. In 1951 the long-range search radar was joined by an AN/CPS-5 height-finder radar. Operations ceased with the activation of site P-12 at North Bend.

L-36 — Fort Stevens (A-1)

An AN/TPS-1B long-range search radar was operational by mid-1950 at this site. In 1951 this radar was augmented with an AN/CPS-5. Site P-57 at Naselle, Washington, assumed coverage of the area in February 1952.

P-12/Z-12 — Reedsport/North Bend (A-4, A-9/GCI)

The 761st AC&W Squadron began operating AN/FPS-3 and AN/FPS-4 radars at this site in February 1952. With site P-12's activation, site L-33 at Portland shut down. In 1955 an AN/FPS-8 was installed. The radar subsequently was converted into an AN/GPS-3 and removed in 1956. In 1957 an AN/FPS-6 took over height-finder duties. An AN/FPS-6B was added in 1959. This site began operating an AN/FPS-7 FD search radar in 1960. In June 1960 this site was integrated into the SAGE system. By 1963 AN/FPS-26A and AN/FPS-90 sets were performing height-finding duties. This site became a TAC facility in 1979.

LP-32/P-32/Z-32 — Condon (A-3, A-5, A-9/SS)

The 636th AC&W Squadron was activated in 1947. By 1952 this unit was operating AN/FPS-3 and AN/FPS-4 sets at Condon. These sets were replaced in 1957 and 1958 with AN/FPS-20 and AN/FPS-6 sets. In 1959 a second height-finder radar came with the installation of an AN/FPS-6A. The site joined the SAGE system in July 1960. The AN/FPS-20 subsequently was upgraded in 1961 and redesignated as an AN/FPS 66. By 1966 an AN/FPS-27 was in operation at this site. The Air Force deactivated the 636th in September 1970.

M-100/Z-100 — Mt. Hebo (A-6, A-9/SS)

The 689th AC&W Squadron began operations with AN/FPS-4, AN/FPS-6 and AN/MPS-11 radars in 1957. In the late 1950s an AN/GPS-3 replaced the AN/FPS-4. This radar subsequently was replaced by an AN/FPS-6B set. By 1963 the 689th was operating AN/FPS-24, AN/FPS-26, AN/FPS-90, and AN/FPS-11 sets. Mt. Hebo became an FAA/ADC joint-use facility around 1966. The 689th was deactivated in June 1979.

M-118/Z-118 — Burns (A-6, A-9/SS)

In 1955 the 634th AC&W Squadron began operating an AN/MPS-7 set at this site. By 1959 this radar had been joined by a pair of AN/FPS-6 height-finder radars. An AN/FPS-7 radar replaced the AN/MPS-7 radar in 1960. In January 1961 the site joined the SAGE system. In 1963 two AN/FPS-90s were performing height-finder duties. By 1966 an AN/FPS-66 search radar was at the site. The 634th was deactivated in September 1970.

SM-149/Z-149 — Baker (A-7)

In 1962 the 821st Radar Squadron (SAGE) began operating AN/FPS-35 and AN/FPS-6 radars. In June 1968 the Air Force deactivated the 821st.

TM-180/Z-180/J-82 — Keno, Klamath Falls (A-8, A-9/NCC)

In 1959 the 827th AC&W Squadron began operating an AN/FPS-20A and a pair of AN/FPS-6A height-finder radars. In June 1960 Keno joined the SAGE system. In 1960 Keno also was performing air traffic control duties for the FAA. In 1961 the search radar was upgraded and redesignated as an AN/FPS-67. In 1963 an AN/FPS-90 replaced one of the AN/FPS-6 height-finder radars. In 1979 the site came under TAC jurisdiction.

Z-345/J-81 — Salem/Dallas

The FAA began operations at this facility in 1959. The Air Force came to the site in 1979 when a height-finder radar tower was constructed. The site was part of the JSS network.

Over-the-Horizon-Backscatter (OTH-B) — Christmas Valley

In the late 1980s the Air Force built an OTH-B radar transmitter at this site. The site became operational in late 1990; however, it was placed on standby status in 1991 due to the declining Soviet threat.

SAGE — Portland

A SAGE defense center at Adair AFS controlled the Portland Air Defense Sector during the 1960s.

Pennsylvania

L-9 — Fort Indiantown Gap (A-1)

A Lashup radar became operational in November 1950 at this site. It remained in operation for less than a year.

L-16 — Connelsville (A-1)

The Connelsville Lashup site started operations in July 1950 using an AN/TPS-1C long-range search radar. Coverage was assumed in November 1951 by P-63 at Blue Knob.

P-30/Z-30 — Mud Pond/Benton (A-2, A-9/GCI)

The 648th AC&W Squadron activated a pair of AN/CPS-6B radars at this site starting in October 1951. The AN/CPS-6B search radars remained active until 1961. In 1958 a pair of AN/FPS-6B radars replaced the AN/CPS-6B height-finder radar. In late 1958 Bentonville began providing data for the SAGE system. In 1961 this site received an AN/FPS-35 search radar, but difficulties prevented it from becoming operational in 1961. It became operational in 1962. This radar and another located in Manassas, Virginia, was used in 1962 as part of a missile detection test. The results revealed that the AN/FPS-35 had marginal value for detecting submarine-launched ballistic missiles. In 1963 the search radar was complemented by AN/FPS-26A and an AN/FPS-6 height-finder radar. By the end of 1963 this was a joint-use site. The 648th was deactivated in June 1975.

RP-62/P-62/Z-62 — Oakdale (A-9/SS)

This site designation and the 662nd AC&W Squadron was transferred to Oakdale from Brookfield, Ohio, in 1959 and 1960. The site had an FAA ARSR-1A search radar providing air traffic control data, as well as a pair of AN/FPS-6B height-finder radars. SAGE system operations started in 1961. An AN/FPS-20 was installed in 1962. In 1963 this radar operated with AN/FPS-24 search radar as well as AN/FPS-26A and AN/FPS-90 height-finder radars. The 662nd was deactivated in December 1969.

P-63 — Blue Knob Park, Claysburg (A-3)

The 772nd began operations at this peak in April 1952, allowing shutdown of the Connelsville L-16 Lashup site. In 1956 an AN/FPS-4 replaced the original AN/CPS-4 height-finder radar and two years later an AN/FPS-20 replaced the original AN/FPS-3 search radar. This site was deactivated in 1961 and repositioned to Gibbsboro, New Jersey (RP-63).

South Carolina

M-113/Z-113 — North Charleston/Charleston AFB (A-6, A-9/SS)

This site became operational in 1955 when the 792nd AC&W Squadron activated an AN/MPS-7 radar. In 1956 operational radars included the AN/MPS-7, AN/GPS-3, AN/MPS-14, and AN/MPS-8 sets. In late 1959 this site was also performing air traffic control duties for the FAA with a newly installed AN/FPS-20A search radar and AN/MPS-14 set. In 1961 the search radar was upgraded and redesignated as an AN/FPS-66 and an AN/FPS-26 height-finder radar was installed. In 1962 North Charleston joined the SAGE system. By 1966 the site was operating an AN/FPS-27. The facility came under TAC jurisdiction in 1979. In 1980 the Air Force built a height-finder tower at Jedburg, an FAA site located twenty-two miles northwest of Charleston and closed Z-113.

SM-159/Z-159 — Aiken (A-7, A-9/SS)

This was the first Phase II Mobile Radar system to achieve operational status. During December 1955 the 861st AC&W Squadron began activating AN/FPS-3, AN/MPS-14, and AN/TPS-10D radars. In 1958 an AN/FPS-20 and an AN/MPS-14 were operational. The site subsequently received an AN/FPS-7C search radar and an AN/FPS-26 height-finder radar. By 1963 these radars were providing data for the SAGE system. The 861st was deactivated in June 1975.

Z-324/J-03 — Jedburg

This FAA-operated JSS site functioned with an AN/FPS-66A radar into the mid-1990s. It had assumed coverage from the North Charleston Z-113 site in the early 1980s.

South Dakota

M-97 — Ellsworth AFB (A-6)

Beneficial occupancy was achieved at this Phase I Mobile Radar site in late 1954 and the 740th AC&W Squadron began operation in 1955 with AN/MPS-7 and AN/MPS-14 radars. In 1959 an AN/FPS-20A replaced the AN/MPS-7 set. The Air Force deactivated the Ellsworth site in 1962.

M-99/Z-99 — Gettysburg (A-6)

Manned by the 903rd AC&W Squadron, Gettysburg became operational in 1956 with AN/MPS-7 and AN/MPS-14 radars. In 1958 an AN/FPS-20A search radar replaced the AN/MPS-7 and an AN/FPS-6A height-finder radar was installed. Gettysburg became a SAGE center in 1959. The search radar subsequently was upgraded and redesignated as an AN/FPS-66 in 1961. In 1963 an AN/FPS-90 was performing height-finder duties. Later in the 1960s an AN/FPS-27 was installed. The Air Force deactivated the 903rd in June of 1968.

SM-134/Z-134 — Pickstown (A-7, A-9/GCI)

The 695th AC&W Squadron began operations in 1961 with an AN/FPS-66 and a pair of AN/FPS-6 radars. The 695th Radar Squadron (SAGE) was deactivated in September 1968.

Tennessee

L-47 — McGhee-Tyson Airport (A-1)

By mid-1950 this Lashup site was operational to provide coverage for nearby Oak Ridge. The AN/TPS-1B radar remained in use until site P-42 at Cross Mountain assumed area radar coverage in May 1952.

P-42 — Cross Mountain, Lake City (A-3)

In June 1952 the 633rd AC&W Squadron began operating a pair of AN/FPS-10 radars from this peak, allowing for the deactivation of L-47 at McGhee-Tyson Airport. An AN/FPS-6 was added in 1958. This site ceased operations on August 1, 1960, due to budget constraints.

SM-144 — Union City (A-7)

In 1958 the 730th AC&W Squadron was operating an AN/FPS-20 radar here. This site was deactivated in 1960 due to budgetary constraints.

SM-145 — Joelton (A-7)

In 1957 the 799th AC&W Squadron began operating AN/MPS-11 and AN/TPS-10D sets here. Eventually these sets were replaced by more modern AN/FPS-6 and AN/FPS-8 sets. Joelton was deactivated in 1960 due to budgetary constraints.

Texas

P-75/Z-75/Z-241 — Lackland AFB (A-4, A-9/NCC)

In 1952 the 741st AC&W Squadron activated an AN/FPS-3 and AN/FPS-4 radar. In 1958 the FPS-4 height-finder radar was replaced by AN/FPS-6 and AN/FPS-6A sets. In late 1959 this station was also performing air traffic control duties for the FAA. At this time the site hosted an AN/FPS-20A radar. By 1966 the facility hosted an AN/FPS-91A radar. The 741st was deactivated in December 1969 and the FAA assumed control. The Houston-based 630th Radar Squadron sent a detachment to this FAA operated site in September 1972 to set up an AN/FPS-6 radar to join the AN/FPS-66 radar already in place.

P-78 — Duncan NAS, Duncanville (A-4, A-9/SS)

In 1952 the 745th AC&W Squadron began operating a pair of AN/FPS-10 radars at this site. In 1958 the height-finder radar was replaced by an AN/FPS-6. This facility closed on July 1, 1964.

RP-78/Z-78 — Perrin AFB

This site became operational in 1962 with an AN/FPS-20 radar. At the end of 1963 it was performing duty as a joint-use facility for the FAA and ADC. The 745th was deactivated in September 1969.

P-79/Z-79/Z-240/J-15 — Ellington AFB (A-4, A-9/GCI)

In April 1952 the 747th AC&W Squadron began operating a pair of AN/FPS-10 radars. In 1955 the Air Force placed an AN/FPS-8 at this site that subsequently became an AN/GPS-3. This set operated until 1960. In 1957 an AN/FPS-6 set replaced the AN/FPS-10 height-finder radar. By 1960 this facility performed air traffic control duties for the FAA with an ARSR-1 radar. The 747th was deactivated in December 1969 and the FAA operated the ARSR-1. In late 1972 a detachment of the 630th Radar Squadron arrived to operate an AN/FPS-6 height-finder radar.

M-88/Z-88 — Amarillo AFB (A-6, A-9/SS)

Beneficial occupancy was achieved by the 688th AC&W Squadron at this Phase I Mobile Radar site in late 1954. The site was operational with AN/MPS-7 and AN/TPS-

10D radars in 1955. In 1958 the site was operating an AN/FPS-20A search radar. An AN/FPS-6A replaced the AN/TPS-10D height-finder radar in 1960. The AN/TPS-10D radar was upgraded and redesignated as an AN/FPS-67 radar in the mid-1960s. Around 1965 this site became an FAA/ADC joint-use site. In September 1968 the Air Force deactivated the 688th.

M-89/Z-89 — Sweetwater (A-6, A-9/GCI)

In 1956 the 683rd AC&W Squadron began operating AN/MPS-11 and AN/TPS-10D at this site. In 1961 an AN/FPS-6A replaced the AN/TPS-10D height-finder radar and the AN/FPS-6A evolved into an AN/FPS-90. In the mid-1960s the search radar was replaced by an AN/FPS-67. The Air Force deactivated the 683rd in September 1969.

TM-186 — Pyote (A-8, A-9/GCI)

In 1958 the 697th AC&W Squadron began operating AN/FPS-3 and AN/FPS-6 radars at this site. In March 1963 the Air Force ordered the site to close. Operation ceased a month later.

TM-187 — Ozona (A-8, A-9/SS)

In 1959 the 732nd AC&W Squadron began operating AN/FPS-3 and AN/FPS-6 radars. Operations ceased in April 1963.

TM-188 — Eagle Pass (A-8, A-9/GCI)

The 733rd AC&W Squadron began operating AN/FPS-20 and AN/FPS-6 radars at this site in 1959. In March 1963 the Air Force ordered the site to close; operations ceased a month later.

TM-189 — Zapata (A-8)

In 1959 the 742nd AC&W Squadron began operating AN/FPS-3A and AN/FPS-6 radars. The site was deactivated the following year due to budgetary constraints.

TM-190 — Port Isabel

This site became operational in 1959. The 811th AC&W Squadron operated AN/FPS-3A and AN/FPS-6 radars. The Air Force deactivated Port Isabel during the following year due to budgetary constraints.

TM-191 — Rockport (A-8, A-9/GCI)

This station became operational in 1959 with AN/FPS-3 and AN/FPS-6 radars. This station, operated by the 813th AC&W Squadron, also performed air traffic control duties for the FAA. Rockport was ordered closed by the Air Force in March 1963.

TM-192 — Killeen (A-8)

In November 1958 the 814th AC&W Squadron began operating AN/FPS-20 and AN/FPS-6 radars. Killeen was deactivated in late 1960 due to budgetary cuts.

TM-193 — Lufkin (A-8)

In January 1959 the 815th AC&W Squadron began operating AN/FPS-3A and AN/FPS-6 radars. The Air Force deactivated Lufkin in 1960 due to budgetary constraints.

Z-228/Z-244/J-27 — El Paso

Using an ARSR-1 radar, this joint-use facility began contributing information to the Air Defense Command network in 1963. In 1972 a detachment of the 4629th Radar Squadron arrived to install and maintain an AN/FPS-6 height-finder radar.

Z-229/Z-243/J-26 — Odessa

Using an ARSR-1 radar, this joint-use facility began contributing information to the Air Defense Command network in 1963. In 1972 the 630th Radar Squadron sent a detachment to set up and operate an AN/FPS-6 height-finder radar.

Z-242/J-16 — Oilton

In 1972 the 630th Radar Squadron sent a detachment to this site to set up and operate an AN/FPS-6 height-finder radar in conjunction with the FAA-operated AN/FPS-66 search set. In 1995 the FAA operated an AN/FPS-67B at this site.

J-26 — Sonora

During the 1980s this FAA radar site was incorporated into the JSS.

PAVE PAWS — Eldorado AFS

Built to provide warning against a submarine-launched ballistic missile attack, the Eldorado site became operational in 1987. This structure consisted of two AN/FPS-115, phased-array radars mounted onto a large triangular building.

Utah

Z-213 — Salt Lake City/Francis Peak

This ARSR-1 joint-use FAA radar became operational in 1962 and supported the ADC radar network.

Z-216 — Cedar City

This ARSR-2 joint-use FAA radar became operational in 1962 and supported the ADC radar network.

Z-240 — Salt Lake City

In the 1960s the Air National Guard activated an AN/FPS-8 radar at this site to support the Air Defense Command network.

VERMONT

L-3 — Fort Ethan Allen (A-1)

This Lashup site became operational in February 1950. The site used an AN/TPS-1B long-range search radar. In September 1951 coverage was assumed by site P-14 at St. Albans.

P-14/Z-14 — Belleview Hill/St. Albans (A-2, A-9/SS)

The 764th AC&W Squadron began operating a pair of AN/CPS-6B radars at this site in September 1951 and assumed coverage provided by L-3 at Fort Ethan Allen. One of these radars remained in service until 1962. The other was retired from service in 1958 with the arrival of an AN/FPS-6. A second AN/FPS-6 unit arrived during 1959. This was the year Belleview Hill joined the SAGE system. Beginning in 1962, this site operated with an AN/FPS-7C search radar. An AN/FPS-26A arrived in 1963 to replace one of the AN/FPS-6 height-finder radars. The 764th was deactivated in June 1979.

M-103 — North Concord, Lyndonville (A-6)

This site became operational in 1956. At that time the 911th AC&W Squadron activated AN/MPS-11 and AN/MPS-14 radars. In 1958 an AN/FPS-6A height-finder radar joined the site. In 1959 the 911th briefly operated an AN/FPS-3 and integrated North Concord into the SAGE system. In March 1963 the Air Force ordered the site to close. Operations ceased on April 30.

VIRGINIA

P-55/Z-55 — Quantico/Manassas (A-3)

The 647th AC&W Squadron began operating AN/FPS-3 and AN/CPS-4 radars in March 1952 and assumed coverage from a Lashup site at Fort Meade. The AN/FPS-3 remained active until 1962. In 1958 AN/FPS-6 and 6B radars took over height-finder chores. The site began feeding the SAGE system in January 1959. This site received an AN/FPS-35 radar in 1961, but problems prevented operations. It became operational in 1962 and was tested to determine if it could detect missile launchings. The radar detected Polaris and Minuteman missile launches from Cape Canaveral on June 28, 1962. Additional tests revealed the radar had marginal value for missile detection. In 1963 an AN/FPS-26A replaced the AN/FPS-6B height-finder radar. The site was removed from service on March 1, 1965.

L-15/LP-56/P-56/Z-56 — Fort Curtis/Cape Charles (A-1, A-3, A-5, A-9/NCC)

This site started in April 1950 as a Lashup site using AN/CPS-5 and AN/CPS-4 radars. This equipment was used when Fort Curtis was incorporated into the permanent network. The unit manning this facility, the 771st AC&W Squadron, continued operating the AN/CPS-4 and as of April 1952 an AN/FPS-3 radar as well. The AN/FPS-3 remained operational until 1962. In 1955 an AN/FPS-8 was installed, converted to an AN/GPS-3, and operated through 1958. By the end of that year, two AN/FPS-6 height-finder radars were activated. During 1959 Cape Charles became a SAGE center. In 1963 the site hosted AN/FPS-7, AN/FPS-6, and AN/FPS-26A radars. In 1963 the site also became an ADC/FAA joint-use facility. The facility came under TAC jurisdiction in 1979.

M-121/Z-121 — Bedford (A-6)

The 649th AC&W Squadron achieved beneficial occupancy at this Phase I Mobile Radar site in late 1954. Operational status was achieved in 1956 with the activation of the AN/MPS-8 and AN/MPS-11 radars. By 1958 the AN/MPS-8 had been superseded by a pair of AN/FPS-6 height-finder radars. In 1959 an AN/FPS-20A search radar replaced the AN/MPS-11 set and Bedford joined the SAGE system. In 1960 this site also began performing air traffic control duties for the FAA. In 1963 the search radar was upgraded and redesignated as an AN/FPS-67. The 649th was deactivated in June 1975.

Z-321/J-01 — Oceana

This JSS site at Virginia Beach operated an AN/FPS-91A radar as of 1995.

SAGE — Fort Lee

This SAGE direction center controlled the Washington Air Defense Sector during the 1960s.

Washington

Arlington

Located near Billingham, this pre-Lashup site was established by the 505th AC&W Group in 1947 to provide ground-control interception training. The site used a World War II vintage AN/CPS-5 radar. On March 27, 1948, General Spaatz ordered the Air Defense Command to place the radar on twenty-four-hour-a-day status to cover the approaches to Hanford.

Hanford

This was a pre-Lashup site. In March 1948 an AN/TPS-1B radar was activated at this location. With the establishment of the Lashup system, this radar was removed from service in 1949.

L-28 — Spokane (A-1)

This was a pre-Lashup site. In early 1948 an AN/TPS-1B radar was activated at this site and placed on twenty-four-hour-a-day status to provide coverage for Hanford. Spokane was incorporated into the Lashup system and an AN/TPS-1C was placed in operation in July 1950. Operations ceased in February 1952 with the establishment of LP-60 at Colville.

L-29 — Larson AFB (A-1)

This Lashup station became operational with an AN/CPS-5 radar in March 1950. In 1951 an AN/CPS-1 and AN/CPS-4 were placed in operation. The site was taken off line in January 1952 with coverage assumed by site P-40 at Othello.

L-30 — Richland (A-1)

Richland started operations in July 1950 using an AN/TPS-1B radar. Operations ceased in January 1952 with the activation of P-40 at Othello.

L-31 — Paine Field (A-1)

Paine Field became operational in October 1950 with AN/CPS-1 search and AN/CPS-4 height-finder radars. Operations ceased the following June when operations commenced at site P-1 at McChord AFB.

L-32 — McChord AFB (A-1)

McChord became operational in June 1950 with AN/CPS-4 and AN/CPS-5 radars. The site ceased operations in June 1951 with the establishment of site P-1 at McChord.

L-34 — Neah Bay (A-1)

A pre-Lashup site established in March 1958, this site featured an AN/TPS-1B radar that guarded the approaches to Hanford. In 1951 the site received AN/CPS-4 and AN/CPS-5 radars. Operations ceased in February 1952 with coverage assumed by site P-44 at Bohokus Peak.

L-35 — Pacific Beach (A-1)

The Pacific Beach Lashup site was operational in October 1950 with an AN/CPS-5 radar. Operations ceased in February 1952 as coverage was assumed by site P-57 at Naselle.

P-1 — McChord AFB (A-2)

McChord was the top-priority site for the permanent network. Occupancy of the buildings on site occurred in the fall of 1950 and the Air Force completed installation of two AN/CPS-6B medium-range search and height-finder radars in February 1951. The Lashup site at McChord was subsequently shut down. Performance of these new radars was deemed inferior to the World War II vintage models and the calibration process delayed operational readiness at this and other sites. The site was deactivated in 1960 and repositioned to Fort Lawton (RP-1). The 635th AC&W Squadron operated the site.

RP-1 — Fort Lawton

The P-1 site was repositioned to this site and by 1960 was a joint-use station with the FAA. This site used an FAA ARSR-1C search radar and two AN/FPS-6A height-finder radars. In March 1963 the Air Force directed the site to close and the 635th AC&W Squadron deactivated.

Washington

LP-6/P-6 — Curlew/Mt. Bonaparte (A-2, A-5)

The 638th AC&W Squadron was activated in May 1950 and began operating an AN/FPS-3 medium-range search radar and an AN/FPS-5 height-finder radar beginning in January 1952. In 1959 the squadron was deactivated and the station was converted to an unmanned gap-filler radar site to support P-60 at Colville.

LP-40/P-40/Z-40 — Saddle Mountain/Othello (A-2, A-5)

The 637th AC&W Squadron began operating an AN/FPS-3 medium-range search radar and an AN/FPS-5 height-finder radar in January 1952, assuming coverage from sites L-29 and L-30. In 1956 the Air Force replaced the height-finder radar with an AN/FPS-6. In 1958 the 637th operated an AN/FPS-20 radar plus an AN/FPS-6A. In July 1960 the site provided data to the regional SAGE center. In 1963 this radar was replaced by an AN/FPS-7C set featuring an ECCM capability. The 637th was deactivated on March 31, 1975.

P-44/Z-44/J-80 — Bohokus Peak/Makah (A-2, A-9/NCC)

Assuming coverage from site L-34 at Neah Bay, the 758th AC&W Squadron started operating an AN/FPS-3 medium-range search radar and an AN/CPS-4 height-finder radar at this site in January 1952. The land for this site was leased from the Makah Indian Reservation. During the next decade this site saw a variety of radars. By 1963 the 758th Radar Squadron operated an AN/FPS-7A search radar and AN/FPS-90 and AN/FPS-26A height-finder radars. The site began feeding the Seattle Air Defense Sector SAGE center in February 1960. In October 1979 the site came under TAC jurisdiction. In 1995 the FAA operated an AN/FPS-91A search set at this site.

P-46/Z-46 — Birch Bay/Blaine (A-3, A-9/SS)

The 757th AC&W Squadron began operating a pair of AN/FPS-10 radars at Blaine in January 1952. In 1959 this squadron switched to operating an AN/FPS-20 search radar and AN/FPS-6 and 6A height-finder radars. The site began providing data to the Seattle Air Defense Sector SAGE Center in February 1960. In the 1960s this site converted to an AN/FPS-27 search unit. The 757th was deactivated in March 1979.

P-57/Z-57 — Naselle (A-2, A-9/GCI)

In December 1951 the 759th AC&W Squadron began operating AN/FPS-3 search and AN/FPS-5 height-finder radars, which allowed for the closing of Lashup sites at Fort Stevens and Pacific Beach. In 1955 an AN/FPS-8 search radar was placed on the site and subsequently converted to and redesignated as an AN/GPS-3. In 1958 the 759th began operating an AN/FPS-20 radar, as well as AN/FPS-6 and 6A height-finder radars. SAGE operations began in February 1960. In 1962 the AN/FPS-20 was upgraded to become an AN/FPS-67. This site was closed on April 1, 1966.

LP-60/P-60 — Colville (A-2, A-5)

The 760th AC&W Squadron began operating the AN/FPS-3 and AN/FPS-5 radars at this site in February 1952. This site took over coverage once provided by site L-28, Spokane. In 1956 the AN/FPS-5 height-finder radar was replaced by the AN/FPS-6 model. In 1958 the 760th operated a newly installed AN/FPS-20 radar. In August 1960 this site was deactivated due to budget constraints. Squadron deactivation occurred in November.

SM-151/Z-151/J-79 — Mica Peak (A-7, A-9/SS)

Manned by the 823rd AC&W Squadron, this radar became operational in 1958 with AN/FPS-20 search and AN/MPS-14 radars. In 1959 the station received an AN/FPS-6A height-finder radar and began performing air traffic control duties for the FAA. In 1962 the search radar was upgraded and redesignated as an AN/FPS-67. A year later, the AN/FPS-6 was upgraded into an AN/FPS-90 search set. The 823rd was deactivated in July 1975. The FAA assumed control of the station in 1979 and the Air Force maintained the height-finder tower until 1988. As of 1995 the FAA operated an AN/FPS-67B set at this site.

SAGE — McChord AFB

A SAGE direction center here controlled the Seattle Defense Sector during the 1960s. A Region Headquarters and Region Combat Center was also located here.

SAGE — Larson AFB

This SAGE direction center served the Spokane Air Defense Sector during the 1960s.

West Virginia

P-43/Z-43 — Guthrie

The 783rd AC&W Squadron began operating AN/FPS-3 and AN/FPS-4 radars from this site in June 1952. In 1958 these units were replaced by AN/FPS-20 and AN/FPS-6 sets. In 1962 the search radar was upgraded and redesignated as an AN/FPS-67. A second AN/FPS-6 height-finder radar was added in 1963. The Air Force deactivated the 703rd in June 1968.

WISCONSIN

P-19/Z-19 — Antigo (A-3, A-9/SS)

The 676th AC&W Squadron began operating AN/FPS-3 and AN/FPS-4 radars from this site in June 1952. At the end of 1958 this site was operating AN/FPS-20 and AN/FPS-6A radars. A second AN/FPS-6A height-finder radar was added in 1959. Antigo joined the SAGE system in 1960. During 1962 an AN/FPS-35 replaced the AN/FPS-20 set. The 676th was deactivated in June 1977.

LP-31/P-31 — Elk Horn, Williams Bay (A-2)

In late 1951 the 755th AC&W Squadron began operating a pair of AN/CPS-6B radars. An AN/FPS-6 height-finder radar was added in 1959. This station ceased operations in January 1960 and was reactivated as a gap-filler site, supporting the P-31 site at Arlington Heights, Illinois.

P-35/Z-35 — East Farmington, Osceola (A-2, A-9/SS)

The 674th AC&W Squadron began operating a pair of AN/CPS-6B radars from this site in December 1951. These radars were retired at the end of the decade as an AN/FPS-7 and two AN/FPS-6A height-finder radars were installed during 1959. Also during 1959, Osceola joined the SAGE system. The pair of height-finder radars were replaced by AN/FPS-90 sets in 1963. The 674th was deactivated in March 1975.

M-106 — Two Creeks (A-6)

Manned by the 700th AC&W Squadron, this site was operational in 1956. Budget cuts in 1957 forced the closure of this station.

SAGE — Traux Field

This direction center at Madison controlled the Chicago Air Defense Sector during the 1960s. A Region Headquarters and Region Combat Center was also located here.

Wyoming

TM-201/Z-201 — Sundance (A-8)

Operational status was delayed at this site to allow testing. The tests were used to determine if atomic energy could be used to produce sufficient power to operate a radar station. The tests succeeded and when the site became operational in 1962, the AN/FPS-7C, AN/FPS-6, and AN/FPS-26 sets became the nation's first nuclear-powered radars. This SAGE feeder station was manned by the 731st Radar Squadron. The Air Force deactivated the 731st in June 1968.

Located within the Wyoming section of the Black Hills National Forest six miles north of the community of Sundance, this site was powered by a nuclear power plant situated on the sloping side of Warren Peak. (Official U.S. Air Force photographs courtesy National Archives.)

Z-216 — Rock Springs

This ARSR-2 joint-use FAA radar became operational in 1962 and supported the ADC radar network.

Z-219 — Lusk

This ARSR-2 joint-use FAA radar began feeding information to the ADC radar network in 1963.

Z-224 — Lovell

This ARSR-2 joint-use FAA radar began feeding information to the ADC radar network in 1963.

Appendix: Notes for Site Listings

The following notes are cited in the headings in Part III, following the site designation and location.

1. This was a Lashup site. In September 1948 the Air Force authorized the Air Defense Command to put thirteen additional stations in operation in the Northeastern United States. These stations were in operation by mid-1949. Additional stations were added across the nation in 1949 and 1950. This temporary system was given the name Lashup to distinguish it from the interim system for which the Air Force was seeking appropriations. A total of forty-four stations were established under this program. These sites were designated "L."

2. This site was one of the first of twenty-four stations of the permanent network. On December 2, 1948, the Air Force directed the Army Corps of Engineers to proceed with construction of this and the other twenty-three sites. The first twenty-four stations of the permanent network were activated over a period covering late 1951. Radars used within the network included the AN/CPS-5 (at five sites), AN/CPS-6B (at twelve sites), and the AN/TPS-1B later replaced by the AN/FPS-3 (at seven sites). These sites were designated "P."

3. This site was one of twenty-eight stations built as part of the second segment of the permanent network. Prompted by the start of the Korean War, on July 11, 1950, the Secretary of the Air Force asked the Secretary of Defense for approval to expedite construction of the second segment of the permanent network. Receiving the Defense Secretary's approval on July 21, the Air Force directed the Corps of Engineers to proceed with construction.

4. This station was part of the last twenty-three stations constructed as part of the permanent network. When completely operational in late 1952, the permanent network had seventy-five sites.

5. Because of difficulties with new production radar equipment, this site initially received radar equipment from a former Lashup site to expedite operational status. Thus these stations received an "LP" designation.

6. This station was part of the planned deployment of forty-four Mobile radar stations. Initially, the purpose of mobile stations was two-fold. The first purpose was to provide protection for six SAC bases. Four sites were required to protect each of the six bases. One site used an AN/MPS-7 heavy radar, while the other three sites would use an AN/TPS-1D. The second puropse of the mobile stations was to support the permanent network. The remaining twenty mobile stations were sited around the perimeter of the country to support the permanent network of seventy-five stations. This deployment had

Searching the Skies: The Legacy of the United States Cold War Defense Radar Program

been projected to be operational by mid-1952. Funding, constant site changes, construction, and equipment delivery delayed deployment.

For example, in January 1952, it was proposed to use the radars to provide a double perimeter around vital areas in the northeast and northwest. Delays in deployment allowed for the newer AN/MPS-11 to be used instead of the AN/TPS-1D. Siting for the initial forty-four sites was completed in late 1952 only to undergo more revisions. Three "M" sites were operational on a Lashup basis at the end of 1952 using old equipment. The first actual deployment of what became known as the Phase I Mobile Radar program occurred at MacDill AFB in December 1954. Eventually, twenty-nine of these stations were deployed in Phase I.

7. This site was initially part of Phase II of the Mobile Radar program. The Air Force approved this expansion of the Mobile Radar program on October 18, 1952. At first Phase II called for thirty-five radars to be installed with thirty-two in the United States and three in Canada. By the end of 1954 the number planned for the United States dropped to twenty-seven. By the end of 1955 this number was further reduced to twenty. Radars in this network were designated "SM."

8. This site came into existence under Phase III of the Mobile Radar program. On October 20, 1953, the Air Defense Command requested a third phase of twenty-five radars be constructed under the Mobile Radar program. The radars were to be positioned mostly along the coasts and in the south to prevent an "end-run" by enemy bombers. The Air Force approved this Phase III request in January 1954. In 1957 the program was reduced in scope to twenty-one stations. Only one station was operational in 1957. Radars in this network were designated "TM."

9. This site was incorporated into BUIC I, a manual back-up interceptor control system implemented in 1962. BUIC I provided limited command and control capability in the event the SAGE system was disabled.

This Abbreviation	Indicates This
NCC	station served as a NORAD Control Center
GCI	station had Ground Control Interception capability
SS	station served only as a surveillance site

BIBLIOGRAPHY

Introduction

The literature on the Cold War radar and command and control is extensive and diverse. Material is available in general reference works, chronologies, selected government studies, congressional hearings, reports, oral histories, monographs, and articles that cover much of the subject matter.

For readers who wish to delve deeper into the Cold War radar and command and control systems, there are record collections stored at government archives and records centers, and to a lesser extent, Air Force History and Property Management Offices. The types of records on file include routine correspondence, detailed technical reports, program planning documents, and program summaries. Unfortunately, these repositories are scattered across the country and are often poorly indexed. Also, much of the material is still classified. However, with the end of the Cold War, more of this material should become available. In addition to government sources, private sector sources also should be consulted.

Given these hurdles, this bibliography seeks to serve two purposes. The first is to identify and briefly describe some of the archival collections that a researcher may find invaluable. The second is to identify some of the existing literature on the Cold War radar and command and control systems. This biography is not comprehensive. It is designed to provide direction to readers desiring more detailed material than presented in this study.

Record Repositories

Record Repositories come in all shapes and sizes ranging from the National Archives and the Federal Records Center to small military history offices and corporate records. The most valuable record collections the author consulted while preparing this book are listed in the following paragraphs.

Since the 1940s, the Air Force has played the predominant role in the defense of the skies and space surrounding the United States. Air Combat Command and Air Force Space Command are successor organizations to the former Air/Aerospace Defense Commands. Consequently, records of the Air/Aerospace Defense Command can be found at command history offices located at Langley AFB, Virginia, and Peterson AFB, Colorado. In addition to serving as the history office of the Air Force Space Command, the Peterson office hosts records of the Canadian-American North American Air Defense Command, the Continental Air Defense Command, and the United States Space Command. Much of the material remains classified. Earlier command histories and studies are becoming available. Located a few miles from the Air Combat Command History Office is the Air Combat Command Property Office. This office hosts property records for the radar sites that remain with Air Force jurisdiction.

The Air Force Historical Research Agency (AFHRA), Maxwell AFB, Alabama, is the Air Force's most extensive archive, housing sixty million pages of records. AFHRA also sponsors an oral history program. The collection is open to the public, and the agency has a staff of archivists and historians available to assist researchers. In preparing this study, the author drew upon AFHRA's extensive collection of unit and oral histories.

The Air Force History Support Office (AFHSO), Naval Station Anacostia, Washington, DC, maintains a small research collection and also hosts microfilm copies of many of the documents held at AFHRA through the mid-1970s. Since many of the reels contain both classified and unclassified documents, access to the unclassified documents is difficult. Consequently, researchers might want to consult the original documents at AFHRA.

The Air Force Museum at Wright-Patterson AFB has a Research Center with thousands of files, mostly aircraft related. The center has folders containing press releases, articles, and histories on the Air Defense Command. Other available folders focus on radar.

Before World War II, the Army Signal Corps and the Navy Research Laboratory made important strides in developing radar. The Army Communications and Electronics Command Museum at Fort Monmouth, New Jersey, maintains some important historical documents and photographs from this era. During the war, the Radiation Laboratory at Massachusetts Institute of Technology became a major center for radar development. In the 1950s, MIT became host to Lincoln Laboratory. The MIT Institute of Archives and Special Collections holds the records of MIT Presidents Karl T. Compton and James R. Killian who played critical roles in support of both of these endeavors.

The Hoover Institute on War, Revolution, and Peace Archives at Stanford University holds the papers of Colonel Oliver W. Miller, USAF. Due to his role in air defense during the late 1940s and early 1950s, Miller's papers include news releases, radio messages, printed matter, photographs, clippings, and maps related to radar development and its use in the defense of the United States and Canada.

The Army Corps of Engineers supervised the construction of hundreds of radar and command and control sites. Construction records and written histories of many of the projects may be found at regional division and district history offices. Many division and district histories and records have been preserved in the Research Collection, Office of History, Headquarters Army Corps of Engineers, (HQCE) Alexandria, Virginia.

The Army once provided a key component of the nation's defense in the form of Nike missile batteries. Although the Army Air Defense Command was disestablished in 1975, records that discuss how these batteries were integrated into the air defense structure can be found at the Army Center for Military History (CMH), Washington, DC; the U.S. Army Military History Institute (MHI), at Carlisle Barracks, Pennsylvania; and at the U.S. Army Air Defense Artillery Center and School at Fort Bliss, Texas.

Beginning in the 1950s, the Federal Aviation Administration and the Department of Defense began to cooperate. Over the years, the FAA has assumed operation of many of the DoD radar sites. The historian within the Department of Transportation's FAA Office of Public Affairs in Washington, DC, maintains records detailing the beginnings of this cooperative effort. Relevant materials can also be found at the Department of Transportation Library.

Defense contractors also may be a source of additional research material. Recent mergers and takeovers may make the paper trail difficult to follow. For example, the defense electronics division of Westinghouse that made radars was recently sold to Grumman Northrop. Lockheed Martin owns the aerospace division and radar production component formerly held by General Electric. Earlier, General Electric purchased RCA, another radar manufacturer. Each of the remaining defense contractors have corporate relations offices that should be able to provide direction.

One location of corporate materials is the Historical Electronics Museum located near the Baltimore-Washington International Airport in Maryland. The museum has a library and archive holding radar-related materials. There is also an extensive collection of interviews with people who contributed to designing and building various systems. Because the museum has a World War II era SCR-270 radar, there is an extensive photo collection covering this system. Another repository of corporate material is located at the Hall of History at Schenectady, New York. The collection includes materials, including photographs, from the General Electric Company and hundreds of publications discussing technical information and product development. The photograph file consists of GE's central photograph file that was discontinued in the 1970s. Within the collection are hundreds of photographs pertaining to GE's military projects.

Finally, many of the individuals who designed the radar and command and control systems were members of the Institute of Electrical and Electronic Engineers. Records of this organization and other resources relating to the profession may be located at the Center for the History of Electrical Engineering located in New Brunswick, New Jersey. The center is particularly helpful as an information clearinghouse, putting researchers in contact with the right people.

Published Sources

Consulting the published sources is always a good first step for any researcher. Unfortunately, little has been written to provide broad overviews of Cold War radar and command and control systems. However several pieces, ranging from scholarly monographs, to magazine articles, do contribute to our understanding of the pieces of the story.

General Reference Works

The following general works provide the reader with an overview of Cold War and military developments that place radar and command and control systems in context. They also identify additional resources.

Encyclopedia of the American Military: Studies of the History, Traditions, Policies, Institutions and Roles of the Armed Forces in War and Peace. Vol. I, II, III. John E. Jessup, ed., New York, NY: Charles Scribner's Sons, 1994.

Finn, Bernard S. *The History of Electrical Technology: An Annotated Bibliography.* New York, NY: Garland Publishing, Inc., 1991.

Futrell, Robert Frank. *Ideas Concepts, Doctrine: Basic Thinking in the United States Air Force.* Maxwell Air Force Base, Montgomery, AL: Air University Press, [1971] 1989.

Gaddis, John Lewis. *Strategies of Containment: A Critical Appraisal of Postwar American National Security Policy.* New York, NY: Oxford University Press, 1982.

Halle, Louis J. *The Cold War as History.* New York, NY: HarperPerennial, 1991.

McCormick, Thomas J. *America's Half Century: United States Foreign Policy in the Cold War.* Baltimore, MD: Johns Hopkins University Press, 1989.

Millet, Alan R., and Peter Maslowski. *For the Common Defense: A Military History of the United States of America.* New York, NY: The Free Press, 1984.

Mueller, Robert. *Air Force Bases.* Vol. I. Washington, DC: Office of Air Force History, 1989.

Paterson, Thomas G. *Meeting the Communist Threat: Truman to Reagan.* New York, NY: Oxford University Press, 1988.

Recent Titles in Electrical History: A Selective Bibliography, 1982–1985. New Brunswick, NJ: Center for the History of Electrical Engineering, 1988.

The U.S. War Machine: An Illustrated Encyclopedia of American Military Equipment and Strategy. New York, NY: Crown Publishers Inc., 1978.

Weigley, Russell F. *The American Way of War: A History of United States Military Strategy and Policy.* New York, NY: 1973.

Weisberger, Bernard A. *Cold War, Cold Peace.* New York, NY: American Heritage Press, 1984.

Chronologies

"A Chronology of Air Defense, 1914–1961," ADC Historical Study No. 19, 1962.

"A Handbook of Aerospace Defense Organization, 1946–1986," Peterson Air Force Base, CO: Air Force Space Command Office of History, 1987.

Schaffel, Kenneth. "Appendix 1. Milestones in U.S. Air Defense to 1960," *The Emerging Shield: The Air Force and the Evolution of Continental Air Defense, 1945–1960.* Washington, DC: Office of Air Force History, 1991.

Studies and Reports

The majority of these reports examine, often in considerable detail, the evolution of specific systems or programs. The majority were written by the military. Air Defense Command (ADC) Studies are available at the Air Force History Office, Bolling AFB, or at the Air Force Historical Research Agency, Maxwell AFB. A forthcoming report being prepared by John F. Hoffecker and Mandy Whorton with the Argonne National Laboratory for the U.S. Air Force Space Command promises to provide even greater context for missile and space detection-related installations.

Bell Laboratories. "ABM Research and Development at Bell Laboratories: Project History, October 1975," Whippany, NJ: Bell Laboratories, 1975.

Benson, James R. "An Archaeological Reconnaissance of the Over-the-Horizon Radar Project Transmitter Site, Buffalo Flat, Christmas Lake Valley, Lake County, Oregon," Seattle, WA: U.S. Army Engineer Seattle District, 1986.

Chang, Ike Yi. "The Rise of Active Element Phased Array Radar," Santa Monica, CA: Rand, 1991.

Charo, Arthur. "Continental Air Defense: A Neglected Dimension of Strategic Defense," for the Center of Science and International Affairs, Harvard University, Lantham, MD: University Press of America Inc., 1990.

Denfield, D. Colt. "The Cold War in Alaska: A Management Plan for Cultural Resources," Alaska: U.S. Army Corps of Engineers Alaska District, 1994.

Hoffecker, John F., and Mandy Whorton. "Historic Properties of the Cold War Era: 21st Space Wing Air Force Space Command," Argonne, IL: Argonne National Laboratory, pending.

"Historical Data of the Aerospace Defense Command, 1946–1973," ADC History Office, 1974.

McMullen, Richard. "Air Defense and National Policy, 1951–1957," ADC Historical Study No. 24, 1965.

_____. "Air Defense and National Policy, 1958–1964," ADC Historical Study No. 26, 1965.

_____. "Interceptor Missiles in Air Defense, 1944–1964." ADC Historical Study No. 30, 1965.

_____. "The Birth of SAGE, 1951–1958." ADC Historical Study No. 33, 1965.

_____. "Radar Programs for Air Defense, 1946–1966," ADC Historical Study No. 34, 1966.

_____. "Command and Control Planning: 1958–1965," ADC Historical Study No. 35, 1965.

_____. "The Aerospace Command and Antibomber Defense, 1946–1972," ADC Historical Study No. 39, 1973.

Temme, Virge Jenkins, David Winkler, and John Lonnquest. "Historical and Architectural Documentation Reports of Finley Air Force Station, Finley, North Dakota," Langley AFB, VA: Headquarters, Air Combat Command, 1995.

_____. "Historical and Architectural Documentation Reports of North Truro Air Force Station, North Truro, Massachusetts," Langley AFB, VA: Headquarters, Air Combat Command, 1995.

_____. "Historical and Architectural Documentation Reports of Gibbsboro Air Force Station, Gibbsboro, New Jersey," Langley AFB, VA: Headquarters, Air Combat Command, 1995.

_____. "Historical and Architectural Documentation Reports of Calumet Air Force Station, Calumet, Michigan," Langley AFB, VA: Headquarters, Air Combat Command, 1995.

_____. "Historical and Architectural Documentation Reports of Keno Air Force Station, Keno Oregon," Langley AFB, VA: Headquarters, Air Combat Command, 1995.

Wolf, Richard I. "The United States Air Force Basic Documents of Roles and Missions," Bolling AFB, Washington, DC: Office of Air Force History, 1987.

"Understanding Soviet Naval Developments," 3rd ed. Washington, DC: Office of the Chief of Naval Operations, 1978.

Command and Unit Histories

Usually found in either command history offices or at military repositories such as the Air Force Historical Research Agency, these histories often hold a wealth of information. At the unit level the histories focus on the day-to-day administration and operation of the organization.

Davis, Harry M. *History of the Signal Corps Development of U.S. Army Radar Equipment: Early Research and Development, 1918–1937 (Part I)*. Washington, DC: Office of the Chief Signal Officer, Signal Corps Historical Section Special Activities Branch, 1944.

Kitchens, James H. III. *A History of the Huntsville Division, U.S. Army Corps of Engineers: 1967–1976*. Huntsville, AL: U.S. Army Engineer Huntsville Division, 1978.

The Federal Engineer Damsites to Missile Sites: A History of the Omaha District U.S. Army Corps of Engineers. Omaha, NE: U.S. Army Engineer District Omaha, 1984.

Congressional Hearings and Reports

Congressional hearings and reports contain a wealth of information on radar and command and control programs.

"House Hearing on the Military Establishment Appropriation Bill for FY 1947," pp. 407, 408, 414.

"Hearings before the Subcommittee of the House Committee on Appropriations on the Military Establishment Appropriation Bill for 1948," pp. 17, 629.

"Hearings of the Subcommittee of the House Armed Services Committee on H.R. 2546," p. 338.

"Hearing on Air Force Appropriations for Fiscal 1952, House Appropriations Committee," pp. 235, 240, 594.

"House Hearing on the Air Force Appropriation Bill for Fiscal 1959," March 5, 1958, pp. 36–37; March 7, 1958, p. 115–17; March 10, 1958, pp. 186–87, 215, 218; March 13, 1958, pp. 382–95; March 19, 1958, pp. 727–30, 744–47.

"House Hearings on Department of Defense Appropriation for Fiscal 1961, Revisions in 1960 and 1961 Air Defense Programs," March 20, 1960, pp. 3, 19–22, 25–37.

"House Hearings on Department of Defense Appropriation for Fiscal 1962, Part 2," March 15, 1961, p. 841, 854–55; March 17, 1961, pp. 939–64, 968–70.

"House Hearings on the Department of Defense Budget for Fiscal 1964. House Armed Services Committee," January 31, 1963, p. 323; February 21 1963, pp. 1148, 1170, 1189–90.

"Continental Air Defense: A Dedicated Force is No Longer Needed," United States General Accounting Office Report to Congressional Committees, May 1994.

Books on Strategic Defenses

Kenneth Schaffel's book is especially valuable as he integrates the radar and command and control development with the deployment of different weapon systems and places this story in the context of the early Cold War period.

Baucom, Donald. *The Origins of SDI, 1944–1983.* Lawrence, KS: University Press of Kansas, 1992.

Eglin, James M. *Air Defense in the Nuclear Age.* New York, NY: Garland Publishing Company, 1988.

Englebardt, Stanley L. *Strategic Defenses.* New York, NY: Thomas Y. Crowell Company, 1966.

Jensen, Owen E. "The Years of Decline: Air Defense from 1960 to 1980," in *Strategic Air Defense.* Stephen J. Cimbala, ed., Wilmington, DE: Scholarly Resources Inc., 1989.

Morenus, Richard. *The DEW Line: Distant Early Warning, The Miracle of America's First Line of Defense.* New York, NY: Rand McNally and Company, 1957.

Schaffel, Kenneth, "The U.S. Air Force's Philosophy of Strategic Defense: A Historical Overview," in *Strategic Air Defense.* Stephen J. Cimbala, ed., Wilmington, DE: Scholarly Resources Inc., 1989.

Schaffel, Kenneth. *The Emerging Shield: the Air Force and the Evolution of Continental Air Defense, 1945–1960.* Washington, DC: Office of Air Force History, 1991.

Terrett, Dulany. *United States Army in World War II: The Signal Corps—The Emergency (To December 1941).* Washington, DC: Office of the Chief of Military History, 1956.

Thompson, George R., Dixie R. Harris, and Pauline Oakes. *United States Army in World War II: The Signal Corps—The Test (December 1941 through July 1943).* Washington, DC: Office of the Chief of Military History, 1957.

Thompson, George R., Dixie R. Harris. *United States Signal Corps in World War II—The Outcome (Mid-1943 through 1945).* Washington, DC: Office of the Chief of Military History, 1966.

Books on Radar History and Technology

This study avoided detailed explanations on radar and command and control technologies. Technical overviews can be found in some of the works listed below. For example, Norman Friedman offers an outstanding overview of the evolution of radar technology. Although focused on naval radar systems, Friedman's explanations are applicable to land-based systems. *Tracking the History of Radar* published in 1994 by the Center for the History of Electrical Engineering, Institute of Electrical and Electronics Engineering, provides a much more extensive bibliography of books and articles and identifies repositories of oral histories and other unpublished works related to this subject area.

Blumtritt, Oskar, Hartmut Petzold, and Willaim Asprey, eds. *Tracking the History of Radar.* Piscataway, NJ: Center for the History of Electrical Engineering Institute of Electrical and Electronics Engineers, Inc. 1994.

Blake, Bernard. ed. *Jane's Radar and Electronic Systems*. 6th edition, Alexandria, VA: Jane's Information Group, Inc. 1994.

Brookner, Eli. *Radar Technology*. Boston, MA: Artech House, 1985.

Bryon, Eddie. *Radar: Principles, Technology and Applications*. Englewood Cliffs, NJ: Prentice Hall, 1993.

Cole, Henry W. *Understanding Radar.* 2nd ed. Cambridge, MA: Blackwell Scientific Publications, 1992.

Devereux, Tony. *Messenger Gods of Battle: Radio, Radar, Sonar—The Story of Electronics at War.* London, UK: Brassey's, 1991.

Dunlap, Orrin E., Jr. *Radar: What Radar Is and How It Works*. New York, NY: Harper and Brothers, 1946.

Fagen, M. D., ed. *A History of Engineering and Science in the Bell System: National Service in War and Peace (1925–1975)*. New York, NY: Bell Telephone Laboratories, 1978.

Fisher, David E. *A Race on the Edge of Time: Radar—The Decisive Weapon of World War II*. New York, NY: McGraw-Hill, 1988.

Freeman, Eva C. *MIT Lincoln Laboratory: Technology in the National Interest*. Lexington, MA: Lincoln Laboratory, Massachusetts Institute of Technology, 1995.

Friedman, Norman. *Naval Radar.* Annapolis, MD: Naval Institute Press, 1981.

Kingsley, Simon. *Understanding Radar Systems*. New York, NY: McGraw-Hill, 1992.

Kolosov, Andrei A. *Over the Horizon Radar.* Boston, MA: Artech House, 1987.

Pollard, Ernest C. *Radition: One Story of the MIT Radiation Laboratory, 1940–1945*. Durham, NC: Woodburn, 1982.

Price, Alfred. *The History of US Electronic Warfare: The Renaissance Years, 1946 to 1964*. Washington, DC: Association of Old Crows, 1989.

Skolnik, Merrill I. *Radar Handbook*. New York: McGraw-Hill Book Company, 1970.

Swords, Sean S. *Technical History of the Beginnings of Radar.* London, UK: Peregrinus, 1986.

Toomay, John C. *Radar Principles for the Non-Specialist*. 2nd. ed. New York, NY: Van Nostrand Reinhold, 1989.

Articles

The following list contains articles covering aerospace defense and technological advances.

"Air Defense Command: A Special Report." *Air Force,* vol. 39, no. 6, June 1956, pp. 45–97.

Atkinson, Rick. "Air Defense for Continental U.S. Is Coming Back Into Vogue." *The Washington Post,* August 25, 1984.

Barton, David K. "A Half Century of Radar." *IEEE Transactions on Microwave Theory and Techniques,* MTT-32, No. 9 (Sept. 1984), pp. 1161–70.

Bashow, David L. "The Changing Face of NORAD." *Naval Institute Proceedings,* vol. 121, no. 11, Nov. 1995, pp. 61–63.

Beamer, Samuel C. "Nerve Center for Space Defense." *Air University Review.* vol. 24, Sept.–Oct. 1973, pp. 66–82.

Bell, Alexander Graham. "Preparedness for Aerial Defense." *Air Power Historian,* vol. 2, no. 4, Oct. 1955, pp. 83–87.

Bergquist, Kenneth P. "Parry the Blow and Fight Back." *Air Force,* vol. 38, no. 4, Apr. 1955, pp. 84–86.

Berkner, Lloyd V. "Continental Defense." *Current History,* vol. 26, no. 153, May 1954, pp. 257–62.

Chidlaw, Benjamin W. "Continental Air Defense." *Ordnance,* vol. 39, no. 209, Mar.–Apr. 1955, pp. 706–10.

Corddry, Charles, "How we're building the world's biggest Burglar Alarm." *Air Force,* June 1956, pp. 77–81.

Delear, Frank J. "Tireless Sentries in the Sky." *Bee-Hive,* vol. 32, no. 1, Jan. 1957, pp. 22–26.

Kaysen, Carl. "The Vulnerability of the United States to Enemy Attack." *World Politics,* vol. 6, Jan. 1954, pp 190–208.

Key, William G. "Air Defense of the United States." *Pegasus,* vol. 19, no. 5. Nov. 1952, pp. 1–6.

Killian, James R., and A. G. Hill. "For a Continental Defense." *Atlantic,* vol. 192, no. 5, Nov. 1952, pp. 37–41.

Klass, Philip J. "DEW Line Demonstrates Effectiveness." *Aviation Week,* Aug. 31, 1959.

Kuter, Laurence S. "The Gaps in Our Aerospace Defense." *Air Force,* Aug. 1962, pp. 47–50.

La Fay, Howard, "DEW Line: Sentry of the Far North." *National Geographic,* vol. 114, no.1, Jul. 1958, pp. 128–45.

Lee, Asher. "Trends in Aerial Defense." *World Politics,* vol. 7, Jan. 1955, pp. 233–54.

Miller, Ed. M. "How Air Defense Command Builds a Wall Twelve Miles High." *Air Force,* vol. 39, no. 6, June 1956, pp. 46–47.

Murphy, Charles J.V. "The U.S. as a Bombing Target." *Fortune,* vol. 48, no. 5, Nov. 1953, pp. 118–21.

Partridge, Earle E. "Active Air Defense." *Ordnance,* vol. 43, no. 231, Nov.–Dec. 1958, pp. 386–88.

"SAGE: The New Aerial Defense System of the United States." *Military Engineer,* vol. 48, no. 322, Mar.–Apr. 1956, pp. 115–17.

Saville, Gordon P. "The Air Defense Dilemma." *Air Force,* vol. 36, no. 3, Mar. 1953, pp. 30–33.

Silcox, Marilyn. "Southeast ROCC Marks Beginning of New Air Defense Era," *National Defense,* Jul.–Aug. 1984.

Skolnik, Merrill I. "Fifty Years of Radar." *IEEE Proceedings,* no. 73, Feb. 1985, pp. 182–97.

Smith, Frederic H., Jr. "How Air Defense is Part of the Great Deterrent." *Air Force,* vol. 39, no. 6, June 1956, pp. 90–91, 93.

Strum, Thomas A. "American Air Defense: The Decision to Proceed." *Aerospace Historian,* vol. 19, no. 4, Winter, 1972, pp. 188–94.

"The Truth About Our Air Defense." *Air Force,* May 1953, pp. 25–35.

Watson-Watt, Robert. "Radar Defense Today—and Tomorrow." *Foreign Affairs,* vol. 32, Jan. 1954, pp. 230–43.

White, Thomas D. "We Cannot Have Complete Protection Here at Home." *U.S. Air Services,* vol. 38, no. 12, Dec. 1953, pp. 9–11.

Winchester, James. "The DEW Line Story." *Flying,* vol. 61, no. 2, Feb. 1957, pp. 27–31.

Witze, Claude. "The Gaps in Our Air Defense." *Air Force,* vol. 55, Mar. 1972, pp. 33–39.

Theses and Dissertations

Farquhar, John T. *A Need to Know: The Role of the Air Force Reconnaissance in War Planning, 1941–1953.* Columbus, OH: The Ohio State University, 1991.

Moeller, Steve. *Vigilant and Invincible: The Army's Role in Continental Air Defense, 1950–1974,* Columbus, OH: The Ohio State University, Master's Thesis, 1992.

Defense Radar Acronyms

ABM	Antiballistic Missile
AC&W	Aircraft Control and Warning
ADC	Air Defense Command
	Aerospace Defense Command (1968–1975)
ADCOM	Aerospace Defense Command (1975–1980)
ADIZ	Air Defense Identification Zones
ADTAC	Aerospace Defense Tactical Air Command
AFB	Air Force Base
AFS	Air Force Station
ARSR	Air-Route Surveillance Radar
ASACS	Airborne Surveillance and Control System
AWACS	Airborne Warning and Control System
BMEWS	Ballistic Missile Early Warning System
BUIC	Backup Interceptor Control
CAA	Civil Air Administration
CADS	Continental Air Defense Study
CONAC	Continental Air Command
CONAD	Continental Air Defense Command
DEW	Distant Early Warning
DoD	Department of Defense
ECCM	electronic counter-countermeasure
ECM	electronic countermeasure
FAA	Federal Aviation Administration
FD	frequency-diversity
FY	fiscal year
GATR	ground-to-air transmitter receiver
GCI	Ground Control Interception
GE	General Electric
GOC	Ground Observation Corps
ICBM	Intercontinental Ballistic Missile
JRPG	Joint Radar Planning Group
JSS	Joint Surveillance System

LRRS	Long-Range Radar System
MAR	minimally attended radar
MHz	megahertz
MIT	Massachusetts Institute of Technology
NAS	Naval Air Station
NATO	North Atlantic Treaty Organization
NAVSPASUR	Naval Space Surveillance Command
NORAD	North American Air Defense Command
NSC	National Security Council
OTH-B	Over-The-Horizon-Backscatter
PAR	Perimeter Acquisition Radar
PARCS	Perimeter Acquisition Radar attack Characterization System
PAVE PAWS	Perimeter Acquisition Vehicle Entry Phased-Array Warning System
RAND	Research and Development Corporation
RCA	Radio Corporation of America
RCC	Regional Control Center
ROCC	Region Operation Control Centers
SAC	Strategic Air Command
SAGE	Semi-Automatic Ground Environment
SDI	Strategic Defense Initiative
SLBM	submarine-launched ballistic missile
SPADATS	Space Detection and Tracking System
TAC	Tactical Air Command
UHF	Ultra High Frequency
VHF	Very High Frequency

INDEX

A

ABM Treaty (1972), 47
Aerospace defense, evolution to, 37–56
Aerospace Defense Command (ADCOM), 3, 47, 48, 56
   absorption of, into TAC and SAC, 3, 57
   mission statement of, 48
Airborne Surveillance and Control System (ASACS), 44
Airborne Warning and Control System (AWACS) aircraft, 44, 60
   research and development phase, 47
   for survivable command and control capability, 45–46
Air defense
   building network in, 22–24
   debate over, 24–29
   early development of, 7–14
   improving command and control in, 29–33
   improving radar network in, 33–36
   post-war era in, 14–20, 22
   revitalization of, 57–61
Air Defense Command (ADC), 10, 32
   appropriations battle for, 15–16
   consolidation with TAC into Continental Air Command (CONAC), 19
   formation of, 3
   reestablishment of, 15
Air Defense Command of Canada, 32
Air Defense Identification Zones (ADIZ), 22
Air defense planners, 26
Air defense radar
   deployment of, in 1948, 17
   deployment of, in 1950, 20, 22
Air defense readiness alert, 26
Air Force Communication Service, 48
Air Force Scientific Advisory Board, 24
Air Research and Development Command, 35
Albuquerque (NM) radar station, 17
Alsop, Joseph, 28
Alsop, Stewart, 28
AN/CPS-5 radar, 17, 18
AN/CPS-6 radar, 33
AN/FPS-3 radar, 33
AN/FPS-3A radar, 33
AN/FPS-7 radar, 33
AN/FPS-14 radar, 22, 35
AN/FPS-17 radar, 50
AN/FPS-18 radar, 22, 35–36
AN/FPS-20 radar, 3, 33, 35, 40
AN/FPS-24 radar, 39, 54
AN/FPS-26 radar, 54
AN/FPS-27 radar, 39
AN/FPS-35 radar, 39, 53–54
AN/FPS-49 radar, 49, 50, 54
AN/FPS-50 radar, 49
AN/FPS-64 radar, 3
AN/FPS-65 radar, 3
AN/FPS-66 radar, 3
AN/FPS-67 radar, 3
AN/FPS-80 radar, 50
AN/FPS-85 radar, 53
AN/FPS-91A radar, 3
AN/FPS-92 radar, 50
AN/FPS-108 radar, 50
AN/FPS-115 radar, 55
AN/FPS-118 radar, 60
AN/FSQ-7 radar, 25
AN/FSS-7 radar, 54, 56
AN/GPA-102 radar, 40
AN/GPA-103 radar, 40
Antennas, role of, in development of SCR-268, 10
Antiaircraft Artillery Command, 22
Antiballistic Missile (ABM) system, 37–38

187

AN/TPS-1B/1D radar, 17
Arctic, radar duty in, 68–69
Arlington (WA) radar station, 16
Army Air Forces
 bias of, in favoring offense, 7
 in post-war era, 14–15
Army Corps of Engineers, 23–24
 Omaha District, 45
ARSR-1 radar, 36
ARSR-4 radar, 60
Atomic bomb
 Soviet detonation of, 19
 as threat, 14, 15
Avco Corporation, 54

B

B-10 bombers, 9
B-29 bombers, 15
B-36 program, 19
Backup Interceptor Control (BUIC) I, 41, 44
Backup Interceptor Control (BUIC) II, 44, 45, 46
Backup Interceptor Control (BUIC) III, 47
Backup Interceptor Control (BUIC) system, implementation of, 3, 41, 44
Baker (OR) radar site, 23
Ballistic Missile Early Warning System (BMEWS), 3, 49, 51, 52–53, 63
Barter Island, Alaska, 33
Beale (CA) Air Force Base, 54–55, 56
Bell Telephone Laboratories, 50
Bendix Corporation, 53
Benton (PA) radar site, 53–54
Boeing Company, 19
BOMARC, 7
 closure of missile defense bases, 47
Britain, Battle of, 9, 14
Brown, Harold, 45–46
Brown Plan, 57
Bull, Harold R., 28
Bull Committee, 28
Burroughs Corporation, 44

C

Cambridge Research Laboratory, 25
Canada, extension of American radar network into, 20, 22
Cape Newenham (AK) radar installation, 23
Cavalier (ND) Air Force Station, 52
Charleston (ME) radar site, 54
Chennault, Claire L., 9
Cheyenne Mountain, Colorado, 3, 46, 47, 63
Chiang Kai-shek, 17
China
 Communist forces in, 17
 intervention into Korean War, 23
Christmas Valley (OR) radar site, 61
Clear (AK) radar site, 49–50
Cobra Dane, 50
Cold War, 15
 defensive/offensive strategies during, 7
Columbia (ME) Air Force Station, 60
Continental Air Command (CONAC), 19
Continental Air Defense Command (CONAD), 28, 30, 32
 filter centers, 21
Continental Air Defense Study (CADS), 44
Cuban Missile Crisis, 7, 53
Cudjoe Key (FL) radar site, 57
Czechoslovakia, Communist coup in, 17

D

Defense, U.S. Department of,
 cooperation with Federal Aviation Administration, 36
Distant Early Warning (DEW) Line, 3, 4, 26, 28, 33, 34, 37
Douchet, Giulio, 9
Douglas Aircraft Company's Research and Development (RAND) Project, 16. *See also* RAND Corporation
Duluth International Airport, 47

E

EC-121 Lockheed Constellation planes, 24
Eglin (FL) Air Force Base, 53, 54
Eisenhower, Dwight D., 27, 29, 49, 50
Eldorado (TX) Air Force Station, 55
Electronic counter-countermeasure (ECCM)-modified AN/FPS-20s, 40
Electronic countermeasures (ECM), 33
Electrostatic-storage-tube memory, 25–26
Ent (CO) Air Force Base, 23, 29, 32, 44–45

F

F-12 interceptor program, 46
FAA-operated joint-use sites, 60
Fairchild, Muir S., 18
Federal Aviation Administration, cooperation with Defense Department, 36
Finley (ND) Air Force Station, 24
505th Aircraft Control and Warning (AC&W) Group, 16
Forrestal, James V., 16, 18, 19
Fort Fisher (NC) radar site, 54
Fort Lee (VA) Air Force Station, 47
Frequency-diversity (FD) radar, 35, 39
Fylingdale Moor (England) radar site, 49, 55

G

Gap-filled radar
 ceased operations of, 46
 deployment of, 35–36
Geiger Field (WA) radar site, 23
General Electric, 33, 49
 strike at radar fabrication plant, 22
Gibbsboro Air Force Station, 4
GPA-27, 33
Grand Forks (ND) Air Force Base, 52
Griffiss (NY) Air Force Base, 58
Ground Control Interception (GCI) capability, 9, 21, 44
Ground Observer Corps (GOC), 10, 14, 20, 21
 disestablishment of, 36
 replacement of, 28–29

H

Half Moon Bay (CA) radar station, 16
Hancock Field (Syracuse, NY), 47
Hanford (WA) radar station, 17
Hanscom (MA) Air Force Base, 39, 54
Hebert, F. Edward, 48
Highlands (NJ) radar site, 33
Hiroshima, bombing of, 14

I

IBM, 25
Intercontinental Ballistic Missiles (ICBMs), 26
 introduction of, 37–38

J

Jamming, 33
Johnson, Louis, 19
Johnson, Lyndon B., 47
Joint Radar Planning Group (JRPG), 36
Joint Surveillance System (JSS) project, 4, 57, 60

K

Kelly, Mervin J., 27
Kelly Committee, 27, 28
Kennedy, John F., 44
 and backup for SAGE centers, 41
 and Cuban missile crisis, 7
Killian, James R., 25
Korea, Communist China's intervention in, 23–24
Korean War, 21, 22

L

Lake Charles (LA) radar site, 57
LAMPLIGHT report, 35
Laredo (TX) radar site, 54
Lashup system, 3, 8, 20
LeMay, Curtis, 15
Lincoln Laboratory, 25, 29, 32, 50. *See also* Massachusetts Institute of Technology
Lindbergh, Charles, 19
Luke (AZ) Air Force Base, 47

M

MacDill (FL) Air Force Base, 33, 54, 56
Maginot Line, 27
Magnetic-core memory, 26
Malmstrom (MT) Air Force Base, 47, 52
Manassas (VA) radar site, 53–54
Manhattan Project, 32
March (CA) Air Force Base, 58
Massachusetts Institute of Technology (MIT), 49. *See also* Lincoln Laboratory
  Radiation Laboratory, 12
  Summer Study Group at, 26, 27
McChord (WA) Air Force Base, 23, 47, 58
McGuire (NJ) Air Force Base, 37
McNamara, Robert S., 37–38, 39, 41, 44, 46, 50
Mid-Canada Line, 33
Minuteman missiles, tracking, 54
Minuteman silos, 38
Missile detection and defense, 48–56
Mitchel Field, 10, 15
Mitchell, Billy, 9
Mobile radar deployment, delays in, 23
Montauk (NY) radar station, 17, 39
Moorestown (NJ) radar site, 54
Moscow (ME) Air Force Station, 60
Mountain Home (ID) Air Force Base, 61
Mount Hebo (OR) radar site, 54
Mount Laguna (CA) radar site, 54
Mount Lemmon (AZ) radar site, 47

M-site, 23–24, 33
Myers, Charles T., 22

N

Nagasaki, bombing of, 14
National Security Act (1947), 16
National Security Council (NSC) directive 139, 27
National Security Council (NSC) directive 159/4, 29
National Security Council (NSC) directive 162, 29
Naval Research Laboratory, 9
Naval Space Surveillance System (NAVSPASUR), 50
Neah Bay (WA) radar station, 17
New Look, 29
Nike Ajax antiaircraft missile program, 50
Nike Hercules antiaircraft missile program, 49, 50
Nike missile defense bases, closure of, 47
Nike surface-to-air missiles, 7
Nike X system, 50
Nike Zeus, 49, 50
Nixon, Richard M., 47
North American Air Defense Command (NORAD), 3, 32, 50
  funding cuts in, 39
  operations center for, 54
  regional control center for, 59
North Bay, Ontario, 58
North Truro (MA) Air Force Station, 29, 44, 45
North Warning System, 4
Nuclear deterrence, 63

O

Oceana (VA) Naval Air Station radar site, 57
Operation SKYWATCH, 21

Otis (MA) Air Force Base, 29, 54–55
Over-the-Horizon-Backscatter (OTH-B) radar, 4, 45, 46
   operations center for, 61
   sites for, 47, 60

P

Palermo (NJ) radar station, 17
Partridge, Earle E., 16, 44–45
PAVE PAWS sites, 54–56
Pearl Harbor, bombing of, 9, 10
Perimeter Acquisition Radar Attack Characterization System (PARCS), 50, 52
Pinetree Line, 22, 33
Plexiglas™ plotting boards, 29, 32
Point Arena (CA) radar site, 54, 57
Polaris missiles, tracking, 54
Portland (OR) radar station, 17
Project CHARLES, 25
Project LAMPLIGHT, 35
Project SUPREMACY, 16–19
Project Wizard, 49
P-sites, 23–24
Pulse radar, testing of, 9

R

RADAR (Radio Detection And Ranging), 9
Radar Course Directing Group AN/GSA-51, 44
Radar duty in Arctic, 68–69
Radar network after Sputnik, 39–48
Radar Squadrons (SAGE), 41
Radford, Arthur, 28
Radio Corporation of America (RCA), 49
RAND Corporation, 16
Raytheon, 55
Reagan, Ronald W., 57
Region Operation Control Centers (ROCCs), 4, 57–58

Richmond (FL) Naval Air Station, 36
Robins (GA) Air Force Base, 55

S

Safeguard, 52
Saville, Gordon P., 9, 18–19, 20
SCR-268 radar set, 9, 10
SCR-270 radar set, 9, 10, 11, 12
SCR-271 radar set, 9, 10, 13
Semi-Automatic Ground Environment (SAGE) system, 3, 7, 32, 40
   completion of system, 37
   cuts to command and control program, 39
   direction centers, 41, 44
   Regional Control Centers (RCC), 57
Sentinel, 50
Shaud, John A., 57
Shemya Island (AK) radar site, 50
Siberian Kamchatka peninsula, 50
Signal Corps Laboratories, 9
Sioux City (IA), SAGE direction center at, 41
SM-149 radar, 23
SM-152 radar, 23
SM-sites, 24, 33
Southeast Air Defense Sector, 48
Soviet SS-6 Sapwood ICBM, 41
Soviet Union
   detonation of atomic bomb, 19
   relations with U.S., 17
Spaatz, Carl A., 15–16, 17
Spacetrack System, 50, 53, 54
Spokane (WA) radar station, 17
Sputnik, 41, 48–49, 49
   radar network after, 39–48
SS-6 Sapwood ICBM, 37, 41, 49
Strategic Air Command (SAC), 23, 48, 56
   ADC support for, 15–16
   Air Force support for, 19
   assumption of control of ballistic missile warning system, 48

assumption of spacetracking and
 missile detection functions by, 56
formation of, 3, 15
Strategic Defense Initiative (SDI), 57
Stratemeyer, George E., 15, 16, 19
Submarine-Launched Ballistic Missiles (SLBMs), 53–54
Summer Study Group, 26
 recommendations of, 27

T

Tactical Air Command (TAC), 15, 60
 absorption of ADCOM into, 57
 ADC support for, 15–16
 consolidation with ADC into Continental Air Command (CONAC), 19
 formation of, 3, 15
Thomasville (AL) radar site, 47
Thor-Delta missiles, tracking, 54
Thule, Greenland, radar site, 49, 55
Titan missiles, tracking, 54
TM-sites, 29
Truman, Harry S., 19, 23, 27
Tule Lake (CA) radar site, 61
Twining, Nathan, 28
Twin Lights (NJ) radar station, 17
Tyndall (FL) Air Force Base, 47, 58

U

United States
 military advantage of, over Soviet Union, 7
 relations with Soviet Union, 17
U.S. Naval Forces, 32
Utah Construction, 45

V

Valley, George E., Jr., 24–25
Valley Committee, 24
Vandenberg, Hoyt S., 28
Verona (NY) Test Site, 35

W

Western Electric, 49
Whirlwind II computers, 25, 26, 32, 47, 58
Whitehead, Ennis C., 21
Wilson, Charles E., 28
Wondergem, Casey, 68–69
Woodring, Harry A., 9
World War I, air defense in, 7
World War II, air defense in, 9–14

www.ingramcontent.com/pod-product-compliance
Lightning Source LLC
Chambersburg PA
CBHW080732230426
43665CB00020B/2718